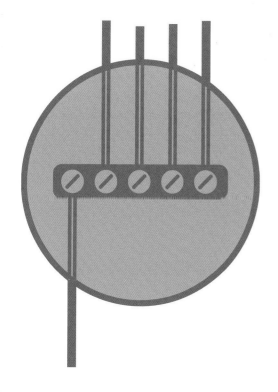

Guidance Note 8
Earthing & Bonding

18th IET Wiring Regulations BS 7671:2018

Published by The Institution of Engineering and Technology, London, United Kingdom

The Institution of Engineering and Technology is registered as a Charity in England & Wales (no. 211014) and Scotland (no. SC038698).

 The Institution of Engineering and Technology is the institution formed by the joining together of the IEE (The Institution of Electrical Engineers) and the IIE (The Institution of Incorporated Engineers).

© 2006, 2012, 2015, 2019 The Institution of Engineering and Technology

First published 2006 (978-0-86341-616-3)
Reprinted 2007
Second edition (incorporating Amendment No. 1 to BS 7671:2008) 2012 (978-1-84919-285-9)
Third edition (incorporating Amendment Nos. 2 and 3 to BS 7671:2008) 2015 (978-1-84919-883-7)
Fourth edition (incorporating BS 7671:2018) 2018 (978-1-78561-413-2)
Reprinted 2019 with minor corrections

Copies of this publication may be obtained from:
The Institution of Engineering and Technology
PO Box 96, Stevenage, SG1 2SD, UK
Tel: +44 (0)1438 767328
Email: sales@theiet.org
www.theiet.org/wiringbooks

ISBN 978-1-78561-413-2 (paperback)

ISBN 978-1-78561-414-9 (electronic)

Typeset in the UK by the Institution of Engineering and Technology, Stevenage
Printed and bound in the UK by Sterling Press Ltd, Kettering

Contents

Cooperating organisations

The Institution of Engineering and Technology acknowledges the invaluable contribution made by the following individuals in the preparation of this Guidance Note.

Institution of Engineering and Technology

M. Coles BEng(Hons) MIET
P. E. Donnachie BSc CEng FIET
Eur Ing G. Kenyon CEng MIET
Eur Ing L. Markwell MSc BSc(Hons) CEng MIET MCIBSE LCGI

We would like to thank the following organisations for their continued support:

BEAMA Installation
Certsure trading as NICEIC and Elecsa
ECA
SELECT
Electrical Safety First (formerly the Electrical Safety Council)
Health and Safety Executive
IET – members of the Users Forum
NAPIT

Guidance Note 8 revised, compiled and edited:

P. E. Donnachie BSc CEng FIET and G Gundry MIET

Acknowledgements

References to British Standards, CENELEC Harmonization Documents and International Electrotechnical Commission (IEC) standards are made with the kind permission of BSI. Complete copies can be obtained by post from:

BSI Customer Services 389 Chiswick High Road London W4 4AL
Tel: +44 (0)20 8996 9000
Fax: +44 (0)20 8996 7001
Email: orders@bsigroup.com

The BSI maintains stocks of international and foreign standards, with many English translations. Up-to-date information on BS standards can be obtained from the BSI website: www.bsi-global.com

Advice is included from Engineering Recommendation G83: *Recommendations for the connection of type-tested small-scale embedded generators (up to 16 A per phase) in parallel with low voltage distribution networks* published by the Energy Networks Association. Complete copies of this and other Engineering Recommendations can be obtained by post from:

Energy Networks Association 4 More London Riverside London SE1 2AU
Tel: 020 7706 5100
Web: www.energynetworks.org

Most of the illustrations within this publication were provided by Farquhar Design: www.farquhardesign.co.uk

Cover design and illustration were created by The Page Design: www.thepagedesign.co.uk

The first edition of this book was written by Eur Ing Geoffrey Stokes BSc(Hons) CEng FIET FCIBSE.

Preface

This Guidance Note is part of a series issued by the Institution of Engineering and Technology to enlarge upon and simplify some of the requirements in BS 7671:2018, the 18th Edition of the Wiring Regulations.

From here on, BS 7671:2018 is referred to as BS 7671. The reference will only include the year after the name of the Standard where there is a need to reference a requirement made in an earlier edition, such as BS 7671:2008.

This Guidance Note does not ensure compliance with BS 7671. It merely explains some of the requirements of BS 7671 relating to earthing, protective bonding and automatic disconnection. Readers should always consult the full text of BS 7671 to satisfy themselves of compliance and must rely upon their own skill and judgement when making use of the guidance provided within this publication.

Electrical installations in the United Kingdom constructed to meet the requirements of BS 7671 are likely to satisfy the relevant aspects of statutory regulations such as the *Electricity at Work Regulations 1989*, but this cannot be guaranteed. It is stressed that it is essential to establish which statutory and other regulations apply and to install accordingly. For example, an installation in premises subject to licensing may have requirements differing from, or additional to, those of BS 7671, and these will take precedence. Users of this Guidance Note should also assure themselves that they have complied with any legislation that post-dates the publication.

Having been constructed to comply with BS 7671, the *Electricity at Work Regulations 1989* impose a duty and responsibility on the owner/user of the electrical installation to maintain it so as to prevent, so far as is reasonably practicable, danger.

Introduction

This Guidance Note is principally concerned with the electrical installation aspects of earthing, protective bonding and the principal protective measure of fault protection against electric shock (automatic disconnection of supply). It draws from the requirements embodied in the relevant parts, chapters and sections of BS 7671.

Earthing and bonding, together with automatic disconnection, are essential aspects of electrical installation design, the requirements for which depend to a large extent on the system type. These aspects are discussed in general terms but, where they apply to special applications, the advice is given in more specific terms, drawing on guidance published in other IET Guidance Notes, in particular Guidance Note 7: *Special Locations*.

The electrical installation designer may find it helpful to refer to the guidance given in this publication, but he/she would be required to take into account all other necessary aspects of the design and, in the process, involve other interested parties, such as:

The Designer
The Installer
The Electricity Distributor
The Installation Owner and/or User
The Architect
The Fire Prevention Officer
All Regulatory Authorities
Licensing Authority
The Health and Safety Executive
The Insurers
The Planning Supervisor.

In producing the design, advice should be sought from the installation owner and/ or user as to the intended use. Often, as in the case of a speculative building, the intended use is unknown. The specification and/or the operational manual must set out the basis of use for which the installation is suitable.

Precise details of each item of equipment should be obtained from the manufacturer and/or supplier and compliance with appropriate standards confirmed. This is particularly important with regard to the protective conductor currents of individual items of current-using equipment and the effect of the accumulation of such currents.

The operational manual should include a description of how the system as installed is to operate and all commissioning records. The manual should also include manufacturers' technical data for all items of switchgear, wiring systems, luminaires, accessories, etc. and any special instructions that may be needed. Section 6 of the Health and Safety at

Work etc. Act 1974 is concerned with the provision of information and instructions on the safe use of such equipment, including any subsequent revisions to that information.

Guidance on the preparation of technical manuals is given in BS EN 82079-1, *Preparation of instructions for use. Structuring, content and presentation. General principles and detailed requirements,* and BS 4940: *Technical information on construction products and services. Guide to content and arrangement.* The size and complexity of the installation will dictate the nature and extent of the manual. This Guidance Note does not attempt to follow the pattern of BS 7671. The order in which the aspects of earthing and bonding as well as automatic disconnection appear is not significant.

Part L of the Building Regulations for England and Wales has specific requirements for building commissioning.

Protective earthing 1

1.1 Protective earthing

Protective earthing is effected by connecting the source, such as the distribution transformer or generator, with Earth and earthing of the fixed and mobile equipment of the electrical installation which it supplies.

1.2 Source earthing

Sometimes referred to as supply system earthing, source earthing is the provision of a connection between the source of energy and the general mass of earth (Earth). This is normally achieved by connecting the neutral point or star point of the secondary winding of the transformer with Earth, via a source earth electrode. This applies whether the source is a distribution transformer or generator. Figure 1.1 illustrates source earthing through a low-impedance source earth electrode; this relates to all TN systems as well as TT systems. For IT systems, the source is connected to Earth via a high impedance or is isolated electrically from Earth.

Source earthing in further detail is beyond the scope of this Guidance Note.

▼ **Figure 1.1** Source earthing

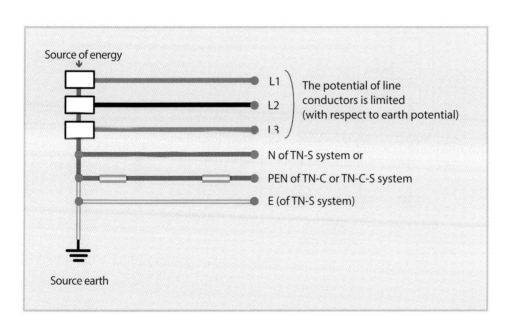

1.3　The purpose of source earthing

The prime purpose of source earthing is to safeguard the security of the supply network by preventing the potential of the live conductors (with respect to Earth) rising to a value that is inconsistent with their insulation rating and which may otherwise lead to a short-circuit and the consequential loss of supply.

For TT systems, the earthing of the source also provides an essential part of the path by which earth fault currents occurring in the installation flow back to the source. These earth fault currents can then be detected by the overcurrent protective devices (such as fuses or circuit-breakers) or residual current devices (RCDs) in order to automatically disconnect the faulty circuit and thus provide fault protection (protection against indirect contact) and, in the case of overcurrent protective devices, protection against fault currents.

For IT systems, the source is connected to Earth via a high impedance or is isolated electrically from Earth.

1.4　The purpose of electrical equipment earthing

Electrical equipment earthing, or electrical installation earthing as it is sometimes referred to, is the connection of all exposed-conductive-parts of the installation to the main earthing terminal, which in turn connects to an appropriate means of earthing, via an earthing conductor, and is always applicable where fault protection is provided by automatic disconnection of supply (ADS). The purpose of such earthing is to facilitate the automatic operation of the protective devices for protection against the effects of earth fault (thermal effects and electric shock), so that the supply to the faulty circuit is disconnected promptly. Figures 1.2 and 1.3 show the earth fault current path of a circuit in which a line-to-earth fault has occurred in the installation, for a TN-C-S system with protective multiple earthing (PME) and for a TT system, respectively.

▼ **Figure 1.2** TN-C-S system with PME – earth fault current path

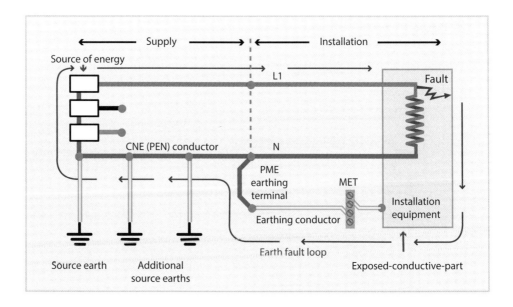

▼ **Figure** 1.3 TT system – earth fault current path

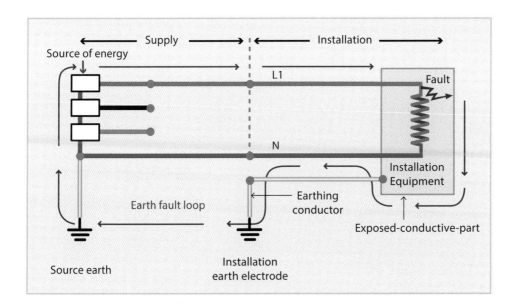

The means of earthing

2

2.1 The means of earthing

The means of earthing is the arrangement that connects the general mass of Earth with the exposed-conductive-parts of an installation via the earthing conductor, the main earthing terminal (MET) and the circuit protective conductors, as shown in Figure 2.1.

▼ **Figure 2.1** The interconnections between the MET and exposed- and extraneous-conductive-parts

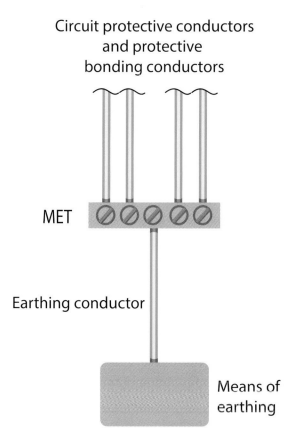

Essentially, the means of earthing can be one of three forms:

▶ the earthing of a source such as a power transformer; or
▶ an earthing facility provided by the electricity distributor to the electricity consumer (e.g. for TN systems); or
▶ an installation earth electrode provided by the consumer (e.g. for TT and IT systems).

542.1.2.1
542.1.2.2
542.1.2.3 For TN-S and TN-C-S systems, the means of earthing is normally provided by the electricity distributor. For TT and IT systems, it falls to the consumer to provide such means of earthing.

For an installation forming part of a TT or IT system, the MET is connected to the installation earth electrode via the earthing conductor.

2.2 Distributors' facilities

542.1.2 For an installation forming part of a TN-S system, the MET is connected via the earthing conductor to the earthed star or neutral point of the source via the electricity distributor's supply cables, either by the cable armouring or sheath or by a separate conductor.

For an installation forming part of a TN-C system, a connection is required to be made between the MET and the electricity distributor's protective earthed-neutral (PEN) conductor, via the earthing conductor.

For an installation forming part of a TN-C-S system, the MET is connected to the earthed point of the source via the installation earthing conductor and the electricity distributor's PEN or combined neutral earth (CNE) conductor.

See also clause 2.15: Responsibility for providing a means of earthing.

2.3 Earth electrodes

It is sometimes necessary to provide the means of earthing for an installation by the use of an installation earth electrode, as in the case of an installation forming part of a TT system. Where an electricity distributor is unwilling or unable to provide a means of earthing in the form of an earth terminal connected to the supply neutral conductor or, if appropriate, the protective conductor of the supply cable, the installation owner is required to provide an installation earth electrode. Similarly, it may, in certain circumstances, be undesirable to employ an electricity distributor's earthing facility; the installation owner is then again required to provide an installation earth electrode.

542.2.2 There are various types of earth electrode (Regulation 542.2.2):

- ▶ driven rods;
- ▶ pipes;
- ▶ tapes;
- ▶ bare wires;
- ▶ plates;
- ▶ underground structural metalwork embedded in foundations;
- ▶ lead sheaths;
- ▶ other metal coverings of cables;
- ▶ suitable metal reinforcement of concrete in reliable contact with earth;
- ▶ structural steelwork of buildings in reliable contact with earth;
- ▶ other suitable underground metalwork; and
- ▶ other suitable metallic water supply pipework (Regulation 542.2.6 refers).

542.2.6 Regulation 542.2.6 of BS 7671 precludes the use of metallic pipes for gases, flammable liquids or a water utility supply as an earth electrode.

Note: Other metallic water supply pipework, such as a private metal supply pipe, may be used as an earth electrode provided precautions are taken against its removal and it has been considered for such use.

2.4 Rod electrodes

Rods are typically constructed of solid copper and are cylindrical in shape. However, rods of copper molecularly bonded to steel are often used too and rods made from galvanised or stainless steel may also be used. Where rigidity is necessary for driving, cruciform or star-shaped sections are sometimes preferred. Although more rigid for driving, these sections do not necessarily provide a significantly lower contact resistance, despite having a larger contact area.

The minimum recommended diameters for rods are generally 9 mm, 12.5 mm and 15 mm for copper and copperclad rods, and 16 mm for rods of galvanized steel and stainless steel. The preferred length of 9 mm diameter rods is 1.2 m and between 1.2 m and 1.5 m for 12.5 mm and 15 mm diameter rods.

Table 10 of BS 7430:2011 *Code of practice for protective earthing of electrical installations* gives guidance as to the minimum cross-sectional area and diameter of various types of electrode, some of which are reproduced here as Table 2.1.

▼ **Table 2.1** Data from Table 10 of BS 7430:2011+A1:2015 *Code of practice for protective earthing of electrical installations*

Electrode type	Cross-sectional area (mm^2)	Diameter or thickness (mm)
Copper tape	50	3
Hard drawn or annealed copper rods or solid wires for driving or laying in ground	50	8
Copperclad or galvanized steel rods (see notes) for harder ground	153	14
Stranded copper	50	3 per strand

Notes:

(a) For copperclad steel rods the core should be of low-carbon steel with a tensile strength of approximately 600 N/mm^2 and a quality not inferior to grade S275 conforming to BS EN 10025. The cladding should be of 99.9 % purity electrolytic copper, molecularly bonded to the steel core. The radial thickness of the copper should be not less than 0.25 mm.

(b) Couplings for copperclad steel rods should be made from copper-silicon alloy or aluminium bronze alloy with a minimum copper content of 75 %.

(c) For galvanized steel rods, steel of grade S275 conforming to BS EN 10025 should be used, the threads being cut before hot-dip galvanizing in accordance with BS EN ISO 1461:2009.

An earth electrode consisting of a driven rod is suitable for providing a means of earthing for many, if not most, installations. However, a driven rod is not appropriate for applications in terrains containing hard strata such as rock.

Where deep driving of rods is necessary, a number of standard lengths of rod can be coupled together, as shown in Figure 2.2. As with driving a single rod, a drive cap should always be used and it is often less problematic to use an electric or pneumatic hammer with suitable rod adaptor to avoid bending of the rod.

▼ **Figure 2.2** Rods coupled together

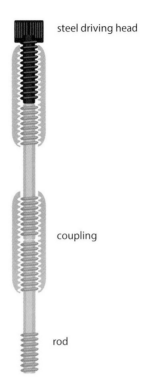

steel driving head

coupling

rod

The position of rod terminations, like all other electrode terminations, is required to be properly identified and accessible for testing and maintenance purposes. Figure 2.3 illustrates a suitable arrangement for termination of the earthing conductor to the rod, consisting of:

► a rod type electrode;
► a rod clamp;
► a warning ('safety') notice; and
► an inspection pit with removable cover (preferably with the words 'Earth rod' embossed into the top surface).

▼ **Figure 2.3** Suitable arrangement for the termination of the earthing conductor

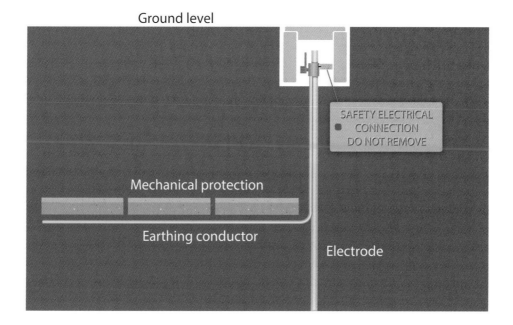

2.5 Tape and wire electrodes

Earth tapes and wires used as an electrode are typically made of untinned copper, strip or round section, and have a minimum cross-sectional area of 50 mm^2 (e.g. 12.5 × 4 mm). The minimum thickness of a tape is 3 mm and the minimum diameter for a strand of a stranded conductor is 3 mm. To avoid adverse soil resistivity conditions created by frosts, these electrodes should be laid at a depth of not less than 1 m. Figure 2.4 illustrates on the left a bare copper tape for use as an electrode buried in the ground, and on the right a suitable colour-identified insulated tape suitable for use as an earthing conductor to connect the electrode with the installation MET.

▼ **Figure 2.4** Bare and insulated copper tape

2.6   Plate electrodes

Earth plates can be made of copper or cast iron. They are typically square, with a surface area of 1 m² to 2 m², and are set vertically at a minimum depth (h) of 600 mm from the ground surface to the top of the plate, to ensure that the soil in close proximity is sufficiently damp. For rock occurring naturally near the surface, this minimum depth requirement may be relaxed. Figure 2.5 illustrates a vertically oriented earth plate and Figure 2.6 shows the connection of the earthing conductor to the plate.

▼ **Figure 2.5** A vertically oriented earth plate

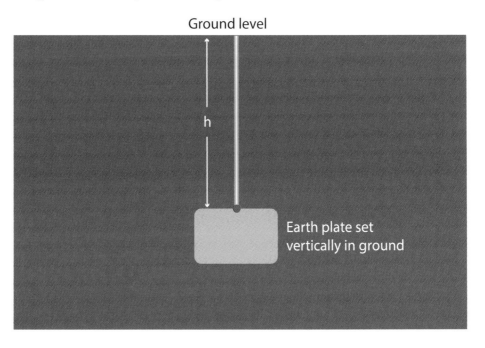

▼ **Figure 2.6** The connection of the earthing conductor to the plate

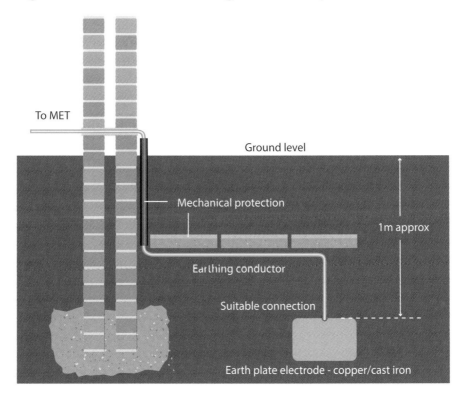

2.7 Structural metalwork electrodes

BS 7430:2011+A1:2015 *Code of practice for protective earthing of electrical installations* recognises that metalwork in foundation concrete can provide a convenient and effective earth electrode. Such electrodes, because of the large electrode area formed by the underground metalwork, can give an overall value of resistance to Earth of well below 1 Ω. However, there are issues that need to be taken into account before deciding on the suitability of using this metalwork as an electrode.

One matter to consider is the possibility of electrolysis and the consequential degradation of the metal. This may occur where the foundation metalwork is electrically in contact with, or bonded to, dissimilar buried metalwork, which may result in corrosion of the metalwork and cracking of the surrounding concrete.

Where significant continuous earth leakage current containing a DC component exists, clause 9.5.8.6 of BS 7430:2011+A1:2015 recommends that an auxiliary earth electrode is bonded to the foundation metalwork.

Where the electrode is made up of structural steel, electrical continuity between all metalwork considered to be part of the earth electrode is essential. For electrical contacts between metalwork within concrete or below ground, such as between reinforcing bars, it is important to ensure this continuity by welding; above ground, joints may be made by attaching a bonding conductor to bypass each structural joint. This particularly applies to surfaces primed with paint before assembly.

Steel stanchions embedded in concrete in the ground may serve as suitable electrodes. Where thus used, it may be necessary to interconnect a number of such stanchions to provide for effective and long-term reliability.

Where the welded metal reinforcement grids in structural concrete are considered for use as an earth electrode, the structural engineer must give prior agreement for their use for such a purpose.

The drilling of structural steelwork for the connection of an earthing conductor, as shown in Figure 2.7, would require the consent of the structural engineer responsible for the design of the building structural components. For this reason, the use of a proprietary earthing and bonding clamp, suitable for the metalwork to which the clamp is to be attached (taking into account the material, cross-section and dimensions), and the protective conductor that is to be terminated on to the clamp (taking into account the material, shape and cross-sectional area) and allowing for the 'looping' in and out of an unbroken protective conductor, may be more beneficial.

▼ **Figure 2.7** Earthing conductor connection to a structural steel stanchion

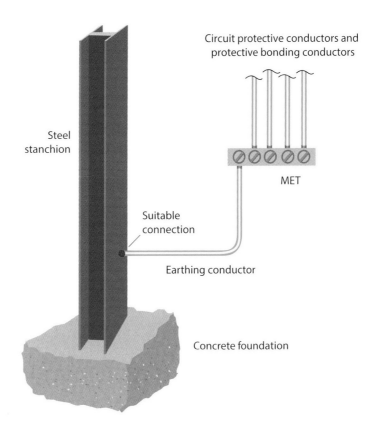

2.8 Lead sheaths and other metal coverings of cables for electrodes

542.2.5 The lead sheath or other metal covering of a cable may be used as an earth electrode subject to the following conditions, as required by Regulation 542.2.5:

▶ adequate precautions are required to be taken to prevent excessive deterioration by corrosion;
▶ the sheath or covering is required to be in effective contact with Earth;
▶ the consent of the owner of the cable is required to be obtained; and
▶ arrangements are required to exist for the owner of the electrical installation to be warned of any proposed change to the cable, which might affect its suitability as an earth electrode.

2.9 Location of the installation earth electrode

542.2.4 The location of the installation earth electrode will depend on a number of factors, which the designer, and possibly the installer, will need to take into account when siting the electrode(s), such as:

▶ the type of electrode
▶ the ground conditions
▶ the level of the water table and its seasonal variation, and
▶ the level at which soil with a suitable conductivity is found.

2.10 Resistance of the earth electrode

The electrical installation designer should determine the maximum acceptable resistance of the electrode as this may influence the type of electrode selected and/ or its location.

542.2.8 In the case of rods and plates, in circumstances in which it may be necessary to achieve a low resistance, a number of rods or plates etc. can be connected in parallel. Dimensions and depths of the electrode, together with the resistivity of the soil, will greatly influence the resistance to Earth.

There are a number of factors which may adversely affect the resistance of an earth electrode, including:

▶ electrode contact resistance;
▶ adjacent soil resistivity;
▶ the size and shape of the earth electrode; and
▶ the burial depth.

The resistance between the electrode and any material in contact with it can be significantly increased by, for example:

▶ corrosion on the electrode;
▶ coatings on the electrode, such as paint;
▶ poor soil (e.g. builder's rubble); and
▶ uncompacted or loose soil surrounding the electrode.

Soil resistivity is defined as a measure of the resistance of a cubic metre of soil, in units of ohm metres (Ωm), and is mainly dependent on soil composition, soil water content and the temperature of the soil. Soil resistivity is given the Greek symbol rho, ρ.

Soil composition can vary enormously. It can be a mixture of sand, soil, chalk and rock, to name just a few examples. Often the choice of position in which to site an electrode is very limited. Some soils are better than others and often the best results are obtained in damp and wet sand, peat clay, arable land, and wet marshy ground.

As Figure 2.8 illustrates, the resistivity can be significantly affected by the soil's moisture content. For example, according to the illustration, when the percentage moisture in the soil is reduced from 20 % to 10 %, the corresponding soil resistivity increases dramatically.

▼ **Figure 2.8** Example of the variation of soil resistivity with the moisture content

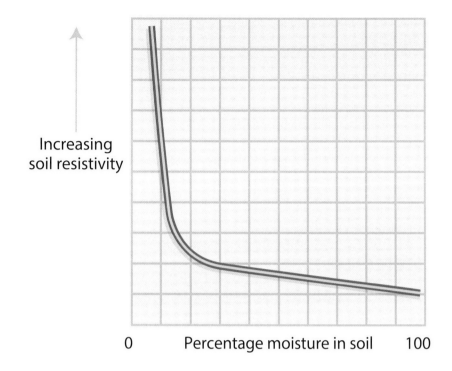

The temperature of the soil does have an effect on the resistivity but this is marginal, except at temperatures around 0 °C (the freezing point of water), where soil resistivity increases disproportionately.

Table 1 of BS 7430:2011+A1:2015 provides data on soil resistivity for general guidance, which is reproduced here for convenience in Table 2.2. Most of the UK would be represented in the second and third columns, but the values given should always be checked against particular local conditions.

▼ **Table 2.2** Soil resistivity (Ωm)

Type of soil	Climatic condition			
	Normal and high rainfall (i.e. greater than 500 mm a year)		**Low rainfall and desert conditions (i.e. less than 250 mm a year)**	**Underground waters (saline)**
	Probable value	Range of values encountered	Range of values encountered	Range of values encountered
Alluvium and lighter clays	5	See note	See note	1 to 5
Clays (excluding alluvium)	10	5 to 20	10 to 100	1 to 5
Marls (e.g. Keuper marl)	20	10 to 30	50 to 300	
Porous limestone (e.g. chalk)	50	30 to 100		
Porous sandstone (e.g. Keuper sandstone and clay shales)	100	30 to 300		
Quartzites, compact and crystalline limestone (e.g. carboniferous sediments, marble, etc.)	300	100 to 1000		
Clay slates and slatey shales	1000	300 to 3000	1000 upward	30 to 100
Granite	1000			
Fissiles, slates, schists, gneiss and igneous rocks	2000	1000 upward		

Note: Depends on water level of locality.

Not surprisingly, the size and shape of the electrode affects the resistance to Earth. Figure 2.9 replicates Figure 3 of BS 7430:1998 *Code of practice for earthing* (which is not included in the 2011 standard) and gives data on the effects of rod diameter and rod length on electrode resistance for soil resistivity of 100 Ωm, assumed uniform.

▼ **Figure 2.9** Variations in rod resistance with size and depth

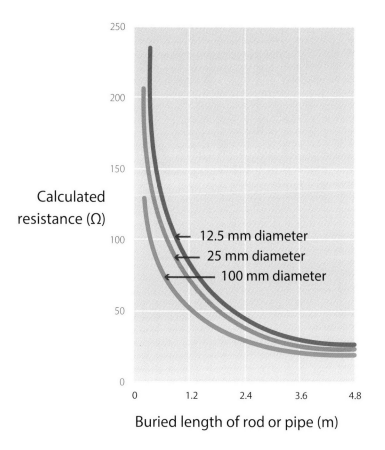

2.10.1 Calculating the resistance of a rod electrode

Where it is necessary to predict the resistance to Earth, R_r, of one vertical rod (or pipe) Equation (2.1) should be used:

$$R_r = \frac{\rho}{2\pi L}\left[\ln\left(\frac{8L}{d}\right) - 1\right](\Omega)$$

(2.1)

where:

L   is the length of the electrode in metres (m).

d   is the diameter of the electrode in metres (m).

ρ is the resistivity of the soil (assumed uniform), in ohm metres (Ωm).

ln   is the natural logarithm (log to the base 'e').

As a practical example of applying Equation (2.1), assume that the limiting value of the electrode resistance R_r is 50 Ω. Checking the resistance for a 15 mm diameter rod electrode buried in porous sandstone ground with a resistivity of 120 Ωm to a covered depth of 1.8 m and inserting the corresponding values into Equation (2.1), gives:

$$R_r = \frac{\rho}{2\pi L}\left[\ln\left(\frac{8L}{d}\right) - 1\right] = \frac{120}{2\pi \times 1.8}\left[\ln\left(\frac{8 \times 1.8}{0.015}\right) - 1\right] = 10.61[(\ln 960) - 1] = 10.61[6.87 - 1] \approx 62\,\Omega$$

(2.2)

Equation (2.2) predicts that a 15 mm diameter rod electrode buried to a depth of 1.8 m will not produce a low enough resistance (we are looking for a maximum of 50 Ω). The diameter of the rod or its length can be changed so as to reduce the predicted resistance value. In this further example, we use a rod of twice the buried length of the original one (i.e. 3.6 m):

$$R = \frac{\rho}{2\pi L}\left[\ln\left(\frac{8L}{d}\right) - 1\right] = \frac{120}{2\pi \times 3.6}\left[\ln\left(\frac{8 \times 3.6}{0.015}\right) - 1\right] = 5.31[(\ln 1920) - 1] = 5.31[7.56 - 1] \approx 35\,\Omega$$

(2.3)

Equation (2.3) gives the result as 35 Ω, which meets our requirements, although two separate rods in parallel and sufficiently spaced may have produced an even better result.

2.10.2 Calculating the resistance of a plate electrode

For an electrode formed by a plate, the resistance to Earth, R, can be predicted by applying Equation (2.4):

$$R = \frac{\rho}{4}\sqrt{\left(\frac{\pi}{A}\right)}\left(\Omega\right)$$

(2.4)

where:

ρ is the resistivity of the soil (assumed uniform) in ohm metres (Ωm).

A is the area of one face of the plate in square metres (m^2).

2.10.3 Calculating the resistance of a tape electrode

For an electrode consisting either of tape or a bare round conductor, the prediction of the resistance to Earth is more problematic, with the results depending on a number of factors in addition to the length and buried surface area, including:

▶ the conductor's shape (round or flat); and
▶ the conductor's arrangement or configuration in the ground.

BS 7430:2011+A1:2015 *Code of practice for protective earthing of electrical installations* gives comprehensive guidance on calculating resistances to Earth for these electrodes. Equation (2.5) gives one such method to predict the resistance, R, of a solitary strip or round conductor run in a single straight line:

$$R_{ta} = \frac{\rho}{2\pi L}\log_e\left(\frac{L^2}{k\,hd}\right)$$

(2.5)

where:

ρ is the resistivity of soil, in ohm metres (Ωm).

L is the length of the strip or conductor, in metres (m).

h is the depth of the electrode, in metres (m).

d is the width of the strip or the diameter of the round conductor, in metres (m).

k has the value 1.36 for strip or 1.83 for round conductor.

2.10.4 Limiting values of the electrode resistance for TT systems

411.5.3 Where the electrode is provided for an installation that forms part of a TT system, and earth fault protection is provided by an RCD, Regulation 411.5.3 requires that Equation (2.6) be satisfied:

$$R_A I_{\Delta n} \leq 50 \ V \tag{2.6}$$

where:

R_A is the sum of the resistances of the earth electrode and the protective conductor(s) connecting it to the exposed-conductive-parts in ohms.

$I_{\Delta n}$ is the rated residual operating current of the RCD (A).

411.3.2.2 It should be noted that the disconnection time must not exceed that given in Regulation
411.3.2.4 411.3.2.2 or 411.3.2.4. (For a system with a nominal voltage U_0 of 230 V, the disconnection time must not exceed 0.2 second or 1 second, respectively.)

Equation (2.6) is generally easily satisfied for 'high-sensitivity' RCDs, such as those having a rated residual operating current not exceeding 30 mA. For example, for an RCD with a rated residual operating current of 30 mA, the limit on R_A would be 1667 Ω, as indicated in Equation (2.7):

$$R_A \leq \frac{50}{30 \times 10^{-3}} \leq 1667 \ \Omega \tag{2.7}$$

Table 41.5 Clearly, satisfying Equation (2.7) is only one point of consideration to take into account when choosing a limit on the earth electrode resistance. The installation designer needs to consider the stability of the electrode over the lifetime of the installation and, for example, the effects of variations in the water table level. BS 7671 recommends that the earth electrode resistance should be as low as practicable: a value exceeding 200 Ω may not be stable (Note 2 to Table 41.5 of BS 7671 refers).

2.11 Electrode installation

542.2.1 The design and construction of an earth electrode requires very careful consideration by the installation designer, as required by Regulation 542.2.1. Examples of issues likely to occur for most, if not all, electrode installations are:

▶ damage – account to be taken of all potential causes of damage to which the earth electrode and earthing conductor could be subjected. Such causes may include, for example, theft, vandalism, gardening, farm equipment, livestock animals, or future excavation associated with a building extension, and will depend to a large extent on the particular circumstances of the location.

▶ corrosion – compatibility with the soil: copper is generally considered to be one of the better and more commonly used materials for earth electrodes. However, the corrosive effects of dissolved salts, organic acids and acid soils should be considered.

▶ corrosion – galvanic effects: this may occur where items of dissimilar buried metalwork are electrically connected together. This electrolytic corrosion can have adverse effects on earth electrodes and earthing conductors, as well as other underground services and structural metalwork. Table 2.3 on next page replicates data given in BS 7430:2011+A1:2015 for the suitability of materials for bonding together.

▼ **Table 2.3** Suitability of materials for bonding together

Material assumed to have larger surface area	Electrode material or item assumed to have the smaller surface area			
	Steel	Galvanized steel	Copper	Tinned copper
Galvanized steel	Suitable	Suitable	Suitable	Suitable
Steel in concrete	Unsuitable	Unsuitable	Suitable	Suitable
Galvanized steel in concrete	Suitable	Suitable[1]	Suitable	Suitable
Lead	Suitable	Suitable[1]	Suitable	Suitable

1 The galvanizing on the smaller surfaces might suffer.

Where it is necessary to use a number of rods or other types of electrode, care should be taken to position the electrodes such that each one is situated outside the resistance area of the others. A commonly adopted principle is to space adjacent electrodes at a distance not less than the buried depth of the electrode, as shown in Figure 2.10, where 's' represents the minimum separation distance between two rods; s being greater than L.

▼ **Figure 2.10** Separation distance for electrodes; s should be not less than L

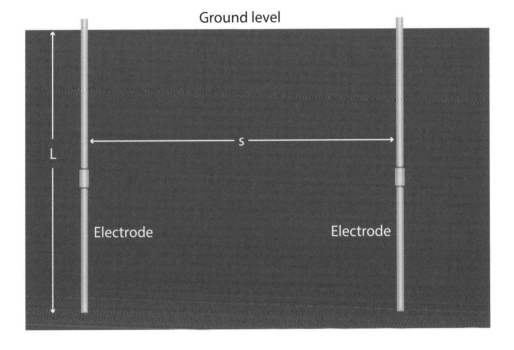

2.12 Electrode loading capacity

The loading capacity of an earth electrode is dependent on:

▶ its shape;
▶ its dimensions;
▶ the current the electrode is required to carry; and
▶ the electrical and thermal properties of the soil.

Essentially, under all operating conditions, the heating effect due to energy dissipated into the soil will not normally result in a detrimental rise in the resistance or failure of the earth electrode. The energy dissipated is I^2t, where 'I' is the electrode current (amps) and 't' is the duration (seconds). *In this consideration, the designer would be required to take account of all currents, not just earth fault currents, likely to flow, such as functional currents.*

For a TT system where fault protection is provided by an RCD, which provides automatic disconnection of the supply, the loading capacity requirement is generally likely to be met routinely. However, the designer still needs to consider electrode load capacity.

2.13 Earth electrode resistance testing

643.7.2 Three methods of measuring the resistance of an earth electrode are described in this section. Test method E1 uses a dedicated earth electrode tester (fall of potential, three- or four-terminal type), test method E2 uses a dedicated earth electrode tester (stakeless or probe type) and test method E3 uses an earth fault loop impedance tester.

Test method E1: Measurement using dedicated earth electrode tester (fall of potential, three- or four-terminal type)

For safety reasons, it is essential to ensure that the entire installation is securely isolated from the supply It is also necessary to disconnect the earthing conductor from the earth electrode. **Caution: If this is the only earth electrode, this may leave the installation unprotected against earth faults. Complete isolation of the installation must be made.** This disconnection will ensure that the test current only passes through the earth electrode and not through any parallel paths. The installation must remain isolated from the supply until all testing has been completed and the earth electrode connection reinstated.

Ideally, the test should be carried out when the ground conditions are least favourable, such as when soil is either frozen or extremely dry.

The test requires the use of two temporary test spikes (electrodes) and is carried out in the following manner:

Connection to the earth electrode is made using terminals C1 and P1 of a four-terminal earth tester. To exclude the resistance of the test leads from the resistance reading, individual leads should be taken from these terminals and connected separately to the electrode. Where the test lead resistance is insignificant, the two terminals may be short-circuited at the test instrument and connection made with a single test lead, the same being true if using a three-terminal tester. Connection to the temporary spikes is made as shown in Figure 2.11.

Figure 2.11 Typical earth electrode test using a three- or four-terminal tester

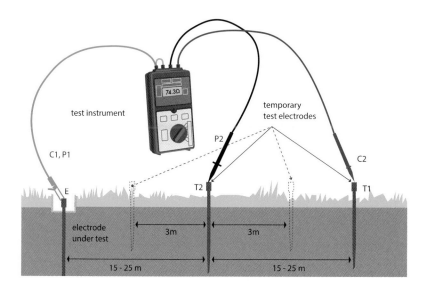

The distance between the test spikes is important. If they are too close together, their resistance areas will overlap. In general, reliable results may be expected if the distance between the electrode under test and the current spike, C2, is at least ten times the maximum dimension of the electrode system, e.g. 30 m for a 3 m long rod electrode.

Three readings are taken:

▶ firstly, with the potential spike, T2, inserted midway between the electrode under test and the current spike, T1;

▶ secondly, with T2 moved to a position 10 % of the overall electrode-to-current spike distance back towards the electrode under test; and

▶ finally, with T2 moved to a position 10 % of the overall distance towards the current spike, from its initial position between the electrode under test and T1.

By comparing the three readings, a percentage deviation can be determined. This is calculated by taking the average of the three readings, finding the maximum deviation of the readings from this average in ohms, and expressing this as a percentage of the average.

The accuracy of the measurement using this technique is typically 1.2 times the percentage deviation of the readings. It is difficult to achieve an accuracy of measurement better than 2 %, and inadvisable to accept readings that differ by more than 5 %. In this event, to improve the accuracy of the measurement, the test must be repeated with a larger separation between the electrode under test and the current spike.

The test instrument output may be AC or reversed DC to overcome electrolytic effects. Because these instruments employ phase-sensitive detectors, the errors associated with stray currents are eliminated.

The instrument should be capable of checking that the resistance of the temporary spikes used for testing is within the accuracy limits stated in the instrument specification. This may be achieved by an indicator provided on the instrument, or the instrument should have a sufficiently high upper range to enable a discrete test to be performed on the spikes.

Where the resistance of the temporary spikes is too high, measures to reduce the resistance will be necessary, such as driving the spikes deeper into the ground or watering with brine to improve the contact resistance. **In no circumstances should the latter technique be used to temporarily reduce the resistance of the earth electrode under test.**

ON COMPLETION OF THE TEST, ENSURE THAT THE EARTHING CONDUCTOR IS RECONNECTED.

Test method E2: Measurement using dedicated stakeless or clamp-based earth electrode tester

Other types of earth electrode resistance tester are available that utilize current transformer (CT) clamps and can carry out measurements without the earth electrode under test having to be disconnected from the installation. The use of two such types is described here.

Instrument using one test coil

This type of test instrument uses a method of measurement similar to the fall of potential method (Method E1, described earlier), in that it uses two temporary test spikes (electrodes), as shown in Figure 2.12. These are placed in the ground, away from the earth electrode under test, in similar fashion to that described for the fall in potential method.

▼ **Figure 2.12** Instrument using one test coil

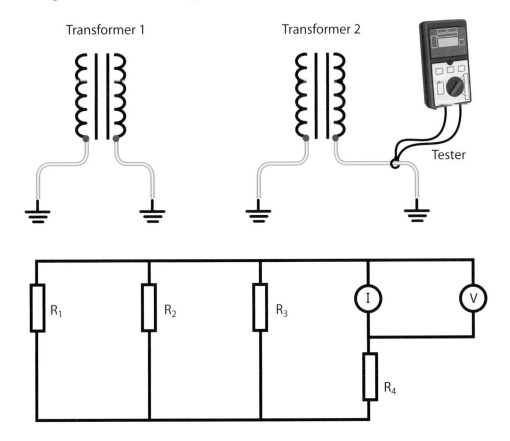

The clamp containing the test coil is placed around the earth electrode under test, or around the conductor connected to that electrode. This eliminates the effects of parallel resistances so that only the resistance earth electrode under test is measured.

The resulting level of accuracy is similar to that given by the fall of potential method.

Instrument using two test coils

This type of test instrument relies for its operation on there being effectively a number of earth electrodes within the installation, not just the electrode under test. The electrodes other than the one under test might not be actual earth electrodes; they might be extraneous-conductive parts buried in the ground or in concrete buried in the ground, such as metallic services pipes or buried structural metalwork.

The instrument uses two coils placed a small distance apart around the earthing conductor of the installation, by means of clamps forming part of the instrument. In practice, the coils may be combined into a single clamp. One coil induces a known voltage in a loop circuit containing the earth electrode under test, the general mass of Earth and other connections with Earth within the installation. The second coil measures the test current.

The instrument carries out a calculation using the formula in Equation (2.8) below. This produces a resistance reading intended to represent the resistance of the earth electrode under test.

$$R_{reading} = R_E + \frac{1}{\frac{1}{R_1} + \frac{1}{R_2} + \frac{1}{R_3} + ... + \frac{1}{R_n}} \qquad (2.8)$$

where:

$R_{reading}$ is the resistance reading given by the test instrument.

R_E is the actual resistance of the earth electrode under test.

R_1, R_2 etc.   are the resistances of the other 'earth electrodes'.

The accuracy of the test reading ($R_{reading}$) depends on the existence of multiple parallel paths for the returning test current to the instrument, such that the effective parallel resistance of these paths is low enough to be neglected.

For example, if there are four other 'earth electrodes', effectively connected in parallel, each having a resistance of, say, 40 Ω, their combined resistance would be 10 Ω. If the resistance of the earth electrode under test was 100 Ω, the total resistance, $R_{reading}$, measured by the test instrument would be 100 Ω +10 Ω =110 Ω. Consequently, the measured value ($R_{reading}$) would be 110 % of the actual value (R_E), an error of 10 %.

However, if there was only one earth electrode other than the one under test, the error in the measurement could be significantly greater, as the effective path would then be through two electrodes effectively connected in series. Using the same values as in the previous example, this would mean the resistance, $R_{reading}$, measured by the test instrument would be 100 Ω + 40 Ω =140 Ω. Consequently, $R_{reading}$ would then be 140 % of the actual value (R_E), an error of 40 %.

Test method E3: Measurement using an earth fault loop impedance tester

643.7.2 note An earth electrode may be tested using an earth fault loop impedance tester. It is recognised that the results may not be as accurate as using a dedicated earth electrode tester.

SWITCH OFF SUPPLY BEFORE DISCONNECTING THE EARTHING CONDUCTOR. The earth fault loop impedance tester is connected between the line conductor at the source of the installation and the earth electrode, and a test performed. The impedance reading taken is treated as the electrode resistance.

▼ **Figure 2.13** Instrument using two test coils

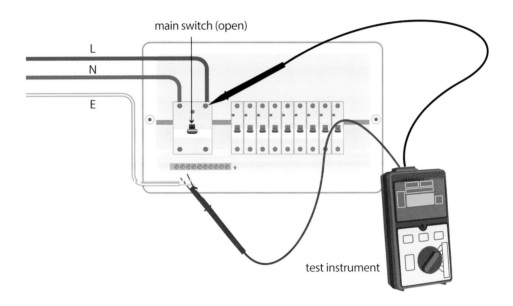

ON COMPLETION OF THE TEST, ENSURE THAT THE EARTHING CONDUCTOR IS RECONNECTED BEFORE THE SUPPLY IS SWITCHED ON AGAIN.

Results of earth electrode testing

For TN-S systems and generator supplies, electrode resistance values may not have been specified, as electrodes often simply provide a local reference earth.

411.5.3 For TT systems, in the absence of the designer's specification, BS 7671 requires RCDs used for fault protection to fulfil the following conditions:

411.5.3 **(a)** The disconnection time shall not exceed that required by Regulation 411.3.2.2 or 411.3.2.4, and
(b) $R_A I_{\Delta n} \leq 50$ V

where:

  ▶ the maximum disconnection time required by Regulation 411.3.2.2 (for final circuits not exceeding 63 A with one or more socket-outlets and 32 A supplying only fixed connected current-using equipment) at a nominal voltage to Earth, U_0, of 230 V is 0.2 s where an RCD is used for fault protection, and

▶ the maximum disconnection time required by Regulation 411.3.2.4 (for a distribution circuit or a circuit not covered by Regulation 411.3.2.2) is 1 s.

R_A is the sum of the resistances of the earth electrode and the protective conductor(s) connecting it to the exposed-conductive-parts (in ohms)

$I_{\Delta n}$ is the rated residual operating current of the RCD (in amps).

The requirements of this regulation are met if the earth fault loop impedance of the circuit protected by the RCD meets the requirements of Table 2.4.

For a nominal voltage, U_0, of 230 V, Table 2.4 gives maximum values of Z_s for non-delayed RCDs, which may be substituted for R_A in the equation in (b) above.

▼ **Table 2.4** Maximum values of earth fault loop impedance (Z_s) for non-delayed RCDs to BS EN 61008-1 and BS EN 61009-1 for U0 of 230 V

RCD rated residual operating current, $I_{\Delta n}$ (mA)	Maximum value of earth fault loop impedance, Z_S (Ω)
30	1667
100	500
300	167
500	100

Table 41.5

Where a time-delayed RCD is used to provide fault protection, the maximum value of earth fault loop impedance including the earth electrode resistance must be such that the requirements of Regulations 411.3 and 411.5 are met. This is likely to require a lower value than given above.

The table indicates that the use of a suitably rated RCD will theoretically allow much higher values of R_A, and therefore of Z_s, than could be expected by using the circuit overcurrent devices for fault protection.

It is advised that earth electrode resistance values above 200 Ω may not be stable, because soil conditions change due to factors such as soil drying and freezing.

2.14 Determination of the external earth fault loop impedance, Z_e

313.1(iv) The external earth fault loop impedance, Z_e, is part of the total earth fault loop impedance and it is essential that an ohmic value for it is determined. There are three ways that the installation designer is permitted to determine this value:

▶ by enquiry to the electricity distributor;
▶ by calculation; and
▶ by measurement.

313.1 (Regulation 313.1 refers.)

Generally, before commencing a detailed design, the installation designer would seek details of the supply characteristics from the electricity distributor, which would include maximum values for prospective fault level and for the external earth fault loop impedance, Z_e.

For the case where a low voltage distribution transformer is owned by the consumer, or where such a transformer is dedicated to supply that customer, the value both for prospective fault level and for the external earth fault loop impedance may be calculated. However, a detailed knowledge of the constituent parts of the loop is required and this can sometimes present difficulties.

Even where the external earth fault loop impedance has been determined, either by enquiry or calculation, it is still necessary for this parameter to be measured in order to ensure that:

▶ the means of earthing is properly connected to the source star or neutral point (either by a metal conductor in the case of TN systems or by an installation earth electrode in the case of TT and IT systems); and
▶ the measured value does not exceed the designer's expected value.

542.4.2
641.4
The measurement of the external earth fault loop impedance, Z_e, is made between the line or phase conductor of the supply and the means of earthing, with the main switch open and with all the circuits securely isolated from the supply. The means of earthing is required to be disconnected from the installation's protective equipotential bonding for the duration of the test, to remove parallel paths. Care should be taken to avoid any shock hazard to the testing personnel and other persons on the site, both whilst establishing connections and performing the test.

ON COMPLETION OF THE TEST, ENSURE THAT THE EARTHING CONDUCTOR IS RECONNECTED BEFORE THE SUPPLY IS SWITCHED ON AGAIN.

(Regulation 542.4.2 refers.)

2.15 Responsibility for providing a means of earthing

Many, if not most, electrical installations are required to be earthed for safety reasons. In fact, under the Electricity Safety, Quality and Continuity Regulations 2002 (ESQCR), the consumer is responsible for ensuring that the installation is satisfactorily earthed. Regulations 26(1) and 26(2) of ESQCR require the consumer's installation to be 'so constructed, installed, protected and used or arranged for use so as to prevent, so far as is reasonably practicable, danger or interference with the distributor's network or with supplies to others'. A consumer's installation complying with the requirements of BS 7671 is deemed to meet these statutory requirements.

In accordance with Regulation 24(4) of the ESQCR, the electricity distributor is generally obliged to provide the consumer with an earthing facility for a new supply to a low voltage installation. The electricity distributor is excused from this obligation if there are safety reasons that preclude the provision of an earthing facility.

Where an existing installation is supplied from the public low voltage network, the electricity distributor is not obliged to provide an earthing terminal. However, the electricity distributor may be willing to do so.

Where the electricity distributor does provide an earthing facility, the responsibility of ensuring the safety and efficacy of this facility rests with the electricity distributor, as required by Regulation 24(1) of the ESQCR. However, it is for the consumer, or his/her agent, to make certain that the earthing facility is suitable for the requirements of the electrical installation and that it is properly connected to the MET of the installation.

Where an earthing facility is not provided by the electricity distributor it should never be assumed that, because a supply cable metal sheath appears to be earthed, such a facility is available. Under no circumstances should the consumer, or his/her agent, connect to the sheath of a supply cable. Such an unauthorized practice is both illegal and dangerous.

GN3 Earth electrodes require maintaining by way of periodic inspection and testing. Guidance Note 3 provides further information.

2

Guidance Note 8: Earthing and Bonding
© The Institution of Engineering and Technology

The earthing conductor

3

3.1 The earthing conductor

The earthing conductor of an electrical installation is the protective conductor that connects the Main Earthing Terminal (MET) of the installation with a means of earthing, as illustrated in Figure 3.1. There is only one such conductor in each installation.

Typically, and for practical considerations, the earthing conductor is provided by:

▶ a single-core cable; or
▶ a core of a multicore cable; or
▶ a tape or strip conductor.

Other types of conductor are permitted by BS 7671, such as:

▶ the metal sheath of a cable; or
▶ the metal armouring of a cable; or
▶ metal conduit.

▼ **Figure 3.1** Conceptual layout of the earthing conductor and the means of earthing

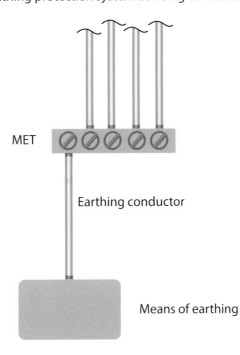

Circuit protective conductors, main protective bonding conductors, functional earthing conductors (if required) and lightning protection system bonding conductor (if any)

MET

Earthing conductor

Means of earthing

542.4.2   A means should be provided to facilitate the disconnection of the earthing conductor in order to measure the external earth fault loop impedance; this should be achieved in an accessible position. The means of disconnection may be in the form of a disconnectable link combined with the MET or bar, as shown in Figure 3.2, or it may be a joint capable of disconnection only by means of a tool (e.g. a spanner or screwdriver).

▼ **Figure 3.2** Illustration of an example of a disconnectable link

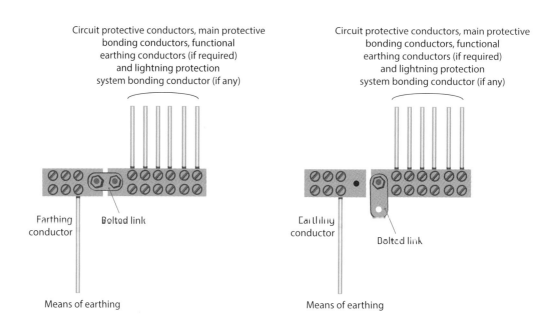

Circuit protective conductors, main protective bonding conductors, functional earthing conductors (if required) and lightning protection system bonding conductor (if any)

Earthing conductor   Bolted link

Means of earthing

Circuit protective conductors, main protective bonding conductors, functional earthing conductors (if required) and lightning protection system bonding conductor (if any)

Earthing conductor   Bolted link

Means of earthing

3.2   The cross-sectional area of an earthing conductor

Subject to certain limits, the minimum cross-sectional area (csa) of an earthing conductor is determined in one of two ways:

543.1.3   ▶   calculated using the adiabatic equation given in Regulation 543.1.3 (reproduced for convenience as Equation (3.1)); or

543.1.4   ▶   by referring to Table 54.7 given in Regulation 543.1.4 (reproduced for convenience
Table 54.7   as Table 3.1).

$$S \geq \frac{\sqrt{(I^2 t)}}{k} (mm^2)$$ (3.1)

where:

S   is the nominal cross-sectional area of the conductor in mm².

434.5.2   I   is the value in amperes (rms for AC) of the fault current for a fault of negligible impedance, which can flow through the associated protective device, due account being taken of the current limiting effect of the circuit impedances and the limiting capability (I²t) of that protective device. Note that account needs to be taken of the effect on the resistance of circuit conductors of their temperature rise as a result of overcurrent (see Regulation 434.5.2).

t   is the operating time of the disconnecting device in seconds corresponding to the fault current I in amperes.

Tables 54.2 to 54.6   k   is a factor that takes account of the resistivity, temperature coefficient and heat capacity of the conductor material, and the appropriate initial and final temperatures. Values of k for protective conductors in various use or service are as given in Tables 54.2 to 54.6. The values are based on the initial and final temperatures indicated within each table.

Where the application of the formula produces a non-standard size, a conductor having the nearest larger standard cross-sectional area should be used.

As an alternative to the use of Equation (3.1) to determine the csa of the earthing conductor, Table 54.7 of BS 7671 gives data for the selection of the csa, reproduced here as Table 3.1.

▼ **Table 3.1** Table 54.7 of BS 7671: Minimum csa of protective conductors in relation to the csa of associated line conductors

csa of line conductor, S	Minimum csa of corresponding protective conductor	
	If protective conductor is of the same material as the line conductor	If protective conductor is not of the same material as the line conductor
(mm²)	(mm²)	(mm²)
$S \le 16$	S	$\frac{K_1}{K_2} \times S$
$16 < S \le 35$	16	$\frac{K_1}{K_2} \times 16$
$S > 35$	$\frac{S}{2}$	$\frac{K_1}{K_2} \times \frac{S}{2}$

Notes:
(a) k_1 is the value of k for the line conductor, selected from Table 43.1 of BS 7671 (replicated in Appendix A) according to the materials of both conductor and insulation.
(b) k_2 is the value of k for the protective conductor, selected from Tables 54.2 to 54.6 as applicable (all except for Table 54.6 are reproduced in Appendix A of this Guidance Note).

543.2.4   As for all protective conductors, an earthing conductor with a csa of 10 mm² or less is required to be of copper (see Regulation 543.2.4).

It has to be acknowledged that the use of Equation (3.1) to determine the csa of an earthing conductor is more difficult than simply referring to Table 54.7 of BS 7671 (reproduced here as Table 3.1). However, using the equation generally gives a more economical size.

Where PME conditions apply

542.3.1
544.1.1
Table 54.8

For an installation for which a PME earthing facility is used as the means of earthing, Regulation 542.3.1 requires the earthing conductor to also meet the csa requirements of Regulation 544.1.1 for main protective bonding conductors. In other words, the csa of the earthing conductor should be able to meet the requirements for a main protective bonding conductor given in Table 54.8 of BS 7671, which for convenience is reproduced here as Table 3.2.

▼ **Table 3.2** Minimum csa of main protective bonding conductors where PME conditions apply

Copper equivalent csa of the PEN conductor (mm^2)	Minimum copper equivalent* csa of the main protective bonding conductor (mm^2)
≤ 35	10
> 35 up to 50	16
> 50 up to 95	25
> 95 up to 150	35
> 150	50

* Copper or copper equivalent (in conductance terms).

It should be noted that the minimum csa of main protective bonding conductors given in Table 54.8 of BS 7671 may be modified by the electricity distributor. In other words, the electricity distributor may require main protective bonding conductors with a larger csa than those given in Table 3.2 owing to, for example, network conditions.

3.3 The csa of a buried earthing conductor

Table 54.1

For an earthing conductor buried in the ground, a further set of minimum csa values is given in Table 54.1 of BS 7671, reproduced for convenience as Table 3.3.

▼ **Table 3.3** Minimum csa of a buried earthing conductor

	Protected against mechanical damage	Not protected against mechanical damage
Protected against corrosion by a sheath	2.5 mm^2 copper 10 mm^2 steel	16 mm^2 copper 16 mm^2 coated steel
Not protected against corrosion	25 mm^2 copper 50 mm^2 steel	25 mm^2 copper 50 mm^2 steel

BS 7430:2011+A1(2015) recommends that earthing conductors should be sufficiently robust to withstand mechanical damage and corrosion.

3.4 Minimum csa of an earthing conductor

543.1.1 Irrespective of the method used to determine the minimum csa of an earthing conductor, Regulation 543.1.1 stipulates that for a copper earthing conductor that is not an integral part of a cable (such as the core of a cable) and is not contained in an enclosure (such as a conduit) formed by a wiring system, the required minimum csa is:

- ▶ 2.5 mm^2 where protection against mechanical damage is provided; or
- ▶ 4 mm^2 where such protection is not provided.

3.5 Impedance contribution of an earthing conductor

Exceptionally, an earthing conductor may be of such length that its csa may need to be increased to minimize the contribution it makes to the overall earth fault loop impedance (Z_s), so as to ensure fault protection for all circuits downstream. This impedance can be reduced by selecting a protective conductor of increased csa or, for an earthing conductor contained within a composite cable, by increasing the csa of all conductors (including that of the line conductors).

3.6 Colour identification of an earthing conductor

514.3.1
514.4.2
If the earthing conductor is a single-core cable or the core of a cable, it must be identified with the bi-colour combination green-and-yellow, as required by Regulations 514.3.1 and 514.4.2. In this combination, one of the colours is required to cover at least 30 % and at most 70 % of the surface being coloured, with the other colour covering the remainder of the surface. See Figure 3.1 for an example of this colour identification scheme in use.

A tape, strip or other bare conductor where used as an earthing conductor must be identified at intervals with the bi-colour combination green-and-yellow, each not less than 15 mm and not more than 100 mm wide, close together, either throughout the length of the conductor or in each compartment and unit and at each accessible position. (See Regulation 514.4.2.)

3.7 Protection of an earthing conductor against external influences

As with all other equipment, an earthing conductor and its electrical connections must be protected from mechanical damage (e.g. impact and vibration), corrosion (e.g. electrolysis) and other external influences to which they may be expected to be exposed. Additionally, an earthing conductor, as for all conductors, must be properly supported throughout its length, particularly at the terminations.

543.3.201 Where an earthing conductor with a csa of 6 mm^2 or less is not part of a multicore cable and not enclosed by conduit or trunking, it is a requirement of Regulation 543.3.201 for it to be protected throughout by a covering at least equivalent to that provided by the insulation of a single-core non-sheathed cable of the appropriate size and having a voltage rating of at least 450/750 V.

Regulation 543.3.201 also requires that an uninsulated earthing conductor with a csa of 6 mm^2 or less, that forms part of a cable, be protected by insulating sleeving (complying with BS EN 60684 series) where the sheath of the cable is removed adjacent to joints and terminations.

543.3.1 As implied by Regulation 543.3.1, in situations in which an earthing conductor is likely to corrode or be subject to mechanical damage, its csa may need to be larger than otherwise determined, in order to protect the conductor against these problems.

521.5.1 However, with regard to mechanical damage, other means of protection may be provided, such as enclosing the earthing conductor in a conduit or in trunking. Where this method is adopted and the conduit or trunking is ferrous, it is important to ensure that the associated live conductors (line conductor(s) and the neutral conductor) are also contained within the same conduit or trunking (see Regulation 521.5.1).

3.8 Disconnection of the earthing conductor

542.4.2 As previously mentioned, Regulation 542.4.2 requires a means for disconnecting the earthing conductor to be provided at or near the MET of an installation. Such a means is necessary to facilitate the measurement of the external earth fault loop impedance, Z_e. Additionally, it is required that the means of disconnection can only be effected by use of a tool. It also needs to be mechanically strong, so as to ensure continuity at all times. Figure 3.2 illustrates a typical means for disconnection of the earthing conductor.

3.9 Connection of the earthing conductor to the means of earthing

Where the means of earthing is provided by the electricity distributor, the connection of the earthing conductor is straightforward and needs no explanation.

However, where the means of earthing is by means of an earth electrode, special care is required to ensure that the electrode is not subject to damage, corrosion or other deterioration. Guidance on these aspects is given in Chapter 2.

The connection of an earthing conductor to an earth electrode is required to be identified by the attachment of a label, as shown in Figure 3.3.

▼ **Figure 3.3** Label to be attached to the point of connection of every earthing conductor to an earth electrode

System types and earthing arrangements

4

4.1 A system

Part 2 In electrical installation terms, a system has two constituent parts, namely a single source, or multiple sources running in parallel, of energy and an electrical installation. The systems are described in a vocabulary in which the various letters have particular meanings, such as those given in Table 4.1. The term *system* is defined in Part 2 of BS 7671.

▼ **Table 4.1** System type designation letters and their meanings

Key

Letter	Designation
T	Terre (French for Earth)
N	Neutral
C	Combined
S	Separate
C-S	Combined then separate
I	Isolated

First letter	Second letter	Subsequent letters
Source earthing arrangements	Arrangement of connection of exposed-conductive-parts of the installation with earth	Arrangement of protective and neutral conductors
T Direct connection of source with Earth at one or more points (e.g. one pole of a single-phase source or the star point of a three-phase source)	N Exposed-conductive-parts of the installation are directly connected by a protective conductor with the source earth	C Single conductor provides both the neutral and protective conductor functions
		S Separate conductors for the neutral and protective conductor functions
		C-S Both the neutral conductor and the protective conductor are combined in the supply and separate in the installation
	T Exposed-conductive-parts of the installation are directly connected by protective conductor(s) to an independent earth electrode and via the conductive mass of Earth to the source earth	N/A

First letter	Second letter	Subsequent letters
I All source live parts isolated from Earth or connected by a high impedance to Earth	T Exposed-conductive-parts of the installation are connected by protective conductor(s) to an independent earth electrode and via the conductive mass of Earth to the source earth	N/A

Note: The ESQCR preclude the IT system for use on public network supplies.

As can be seen from Table 4.1, the first and second letters refer to the earthing arrangements and the others relate to the installation neutral and protective conductor arrangements. There are five distinct system types:

▶ TN-C system;
▶ TN-S system;
▶ TN-C-S system;
▶ TT system; and
▶ IT system.

Sect 312   Before commencing on the detailed design of an electrical installation, it is important for the designer to determine the system type (see Section 312), particularly from the viewpoint of earthing and protective equipotential bonding, as well as fault protection. For example, the system type will have a significant influence on the csa of the earthing and protective bonding conductors.

For the majority of cases, the source is provided by an electricity distributor whose responsibility it is to earth the source (e.g. a distribution transformer). However, for consumers taking the supply at high voltage, and where the distribution transformer is owned by the consumer, the responsibility for earthing the source rests with the consumer. Similarly, it is the consumer's responsibility for earthing the source where the source is a privately owned generator.

Available guidance documents relating to source earthing are:

▶ BS 7430:2011+A1:2015 *Code of practice for protective earthing of electrical installations*, published by the British Standards Institution;
▶ *Guidelines for the design, installation, testing and maintenance of main earthing systems in substations*, originally published by the Electricity Association and now available from the Energy Networks Association; and
▶ The ESQCR (Regulation 8 refers).

4.2   TN-C system

Figure 4.1 shows a TN-C system that consists of a single three-phase source and a three-phase electrical installation with three line conductors (three phase conductors) and a combined neutral and protective conductor throughout. The star point, or neutral if single-phase, of the source is earthed through a low-impedance earth electrode to the general mass of Earth.

▼ **Figure 4.1** TN-C system (three-phase)

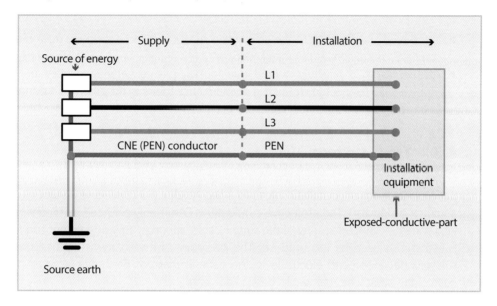

It is important to note that in a TN-C system the neutral and protective earth conductor functions are combined both in the supply and in the installation; in other words, the protective earthed-neutral conductor (PEN conductor, also referred to as a combined neutral and earth (CNE) conductor) is combined throughout the system and the exposed-conductive-parts of the installation are connected by this conductor back to the source. This conductor provides a return path both for the neutral conductor current to flow under normal conditions and for earth fault current to flow for the duration of a line-to-earth fault occurring in the installation. Figure 4.2 shows the complete path that the fault current will take under conditions of a line-to-earth fault occurring in the installation.

▼ **Figure 4.2** TN-C system showing an earth fault loop

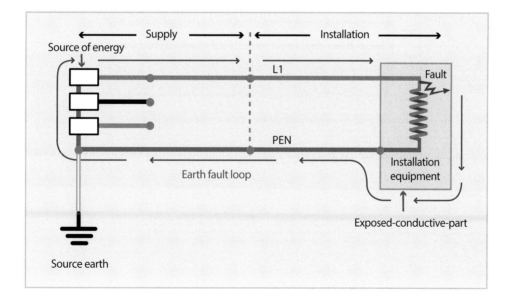

543.4 This system type is rarely used in the UK. Regulation 543.4 of BS 7671 restricts the use of an installation PEN conductor to circumstances where:

▶ as required by the ESQCR, the consumer has obtained an exemption from the ban on CNE conductors in the installation; or

▶ the source is a privately owned transformer, private generating plant or other source and connected in such a manner that there is no electrical connection with the public distribution system.

4.3 TN-S system

Figure 4.3 shows a TN-S system consisting of a single three-phase source and a three-phase electrical installation with four live conductors (three line conductors and a neutral conductor). The protective conductor is commonly provided by means of the lead sheath or wire armouring of the supply cable, although an independent conductor is sometimes used for this purpose. The star point of the source is earthed through a low impedance earth electrode to the general mass of Earth. For a single-phase source, the source neutral would be similarly earthed.

It is important to note that in a TN-S system the neutral conductor and the protective earth conductor functions are separate both in the supply and in the installation. The exposed-conductive-parts of the installation are connected by the protective earth conductor back to the source. This conductor provides a return path for earth fault current to flow for the duration of a line-to-earth fault occurring in the installation. Figure 4.4 shows the complete path that the fault current will take under conditions of a line-to-earth fault occurring in the installation.

▼ **Figure 4.3** TN-S system (three-phase)

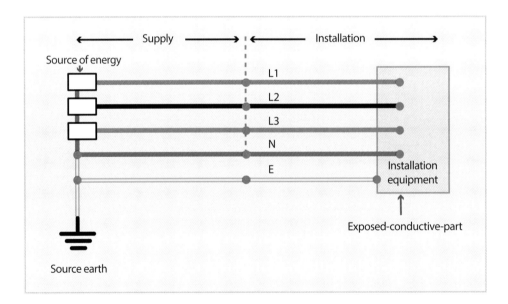

▼ **Figure 4.4** TN-S system showing an earth fault loop

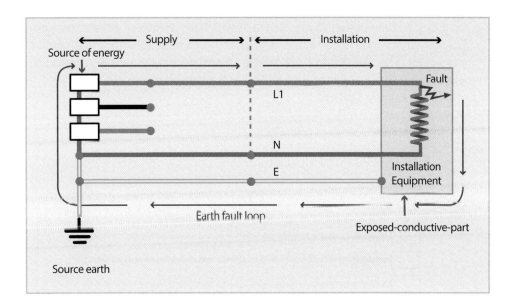

This type of system will be commonly found in many systems supplied from the public supply networks that pre-date the adoption of PME in public electric distribution networks. The electricity distributor often provides an earthing facility connected to the supply cable lead sheath or cable steel wire armouring, or by utilizing a separate supply protective conductor terminated in a suitable earthing terminal located at the supply intake position.

It is worth noting that a supply to an installation forming part of a TN-S system can be provided by an overhead distribution network as well as by the more commonly encountered supply cable buried underground. Figure 4.5 illustrates the two forms of supply.

▼ **Figure 4.5** TN-S system (single-phase) shown pictorially with an overhead and an underground supply cable, respectively

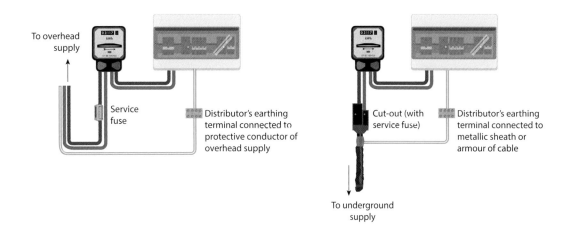

4.4 TN-C-S system

Figure 4.6 shows a TN-C-S system (with protective multiple earthing (PME)), consisting of a single three-phase source and a three-phase electrical installation with three line conductors and a combined neutral and protective (CNE or PEN) conductor in the supply. The star point of the source is earthed through a low-impedance earth electrode to the general mass of Earth. For a single-phase source, the source neutral would be similarly earthed.

▼ **Figure 4.6** TN-C-S system (with PME)

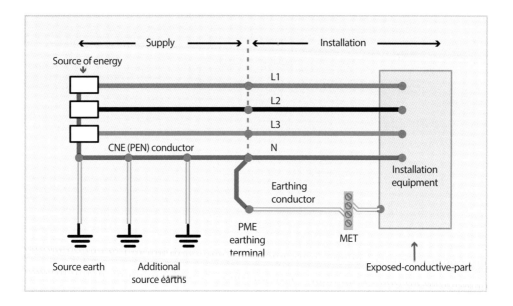

It is important to note that in a TN-C-S system the neutral and protective earth conductor functions are combined in the supply and are separate in the installation. The exposed-conductive-parts of the installation are connected by this separate protective conductor in the installation to the combined neutral and protective conductor of the supply back to the source. This installation protective conductor provides a return path for earth fault current to flow for the duration of a line-to-earth fault occurring in the installation. The combined neutral and protective conductor of the supply provides a return path both for neutral conductor current to flow under normal conditions and for earth fault current to flow for the duration of a line-to-earth fault occurring in the installation. Figure 4.7 shows the complete path that the fault current will take under conditions of a line-to-earth fault occurring in the installation.

▼ **Figure 4.7** TN-C-S system (with PME) showing an earth fault loop

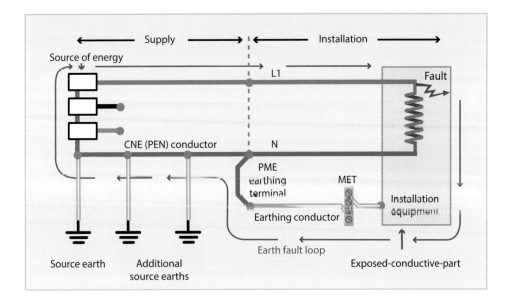

For TN-C-S systems with supplies from the public distribution network, the separation of the functions of the neutral conductor and the protective conductor occurs at the supply intake position. For privately owned sources, the transitional point is normally at the consumer's main switchgear.

There are two basic forms of TN-C-S system, namely:

▶ TN-C-S system with PME (protective multiple earthing); and
▶ TN-C-S system with PNB (protective neutral bonding).

For the TN-C-S PME variant, shown in Figure 4.6, the supply PEN or CNE conductor is earthed at multiple points of the supply, as well as at the source itself, thereby providing a low impedance path to Earth for all parts of the PEN conductor. This system has been used by electricity distributors, and their predecessors, for almost all new low voltage supplies installed since the 1970s.

The use of the TN-C-S PNB variant, shown in Figure 4.8, is confined to cases in which there is only one point in a network at which the consumer's installations are connected to a single source of voltage. The PEN or CNE conductor is connected to Earth at that point, or at another point nearer to the source of voltage.

▼ **Figure 4.8** TN-C-S system with PNB

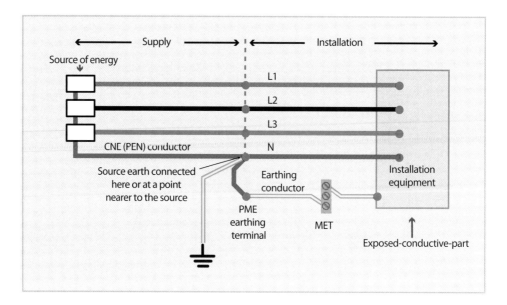

For sources owned by the electricity distributor, the responsibility for earthing the PEN or CNE conductor rests with the distributor. Where the consumer owns the source, it is the consumer's responsibility to earth the PEN or CNE conductor. This is sometimes carried out at a location near the consumer's main switchgear position.

It is worth noting that, as with a TN-S system, the supply to a TN-C-S system (with PME or PND) can be provided by an overhead distribution system as well as the more commonly used supply cable buried underground. Figure 4.9 illustrates the two forms of supply.

▼ **Figure 4.9** TN-C-S system (single-phase) shown pictorially with an overhead and an underground supply cable, respectively

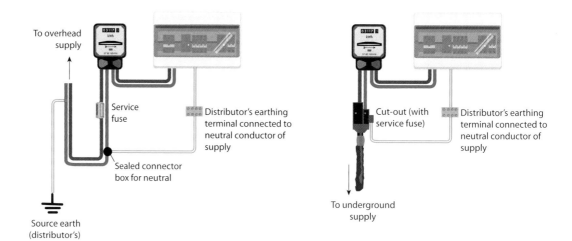

4.5 TT system

Figure 4.10 shows a TT system consisting of a single three-phase source and a three-phase electrical installation with four live conductors (three line conductors and a neutral conductor). The star point of the source is earthed through a low impedance earth electrode to the general mass of Earth. For a single-phase source, the source neutral would be similarly earthed.

▼ **Figure 4.10** TT system

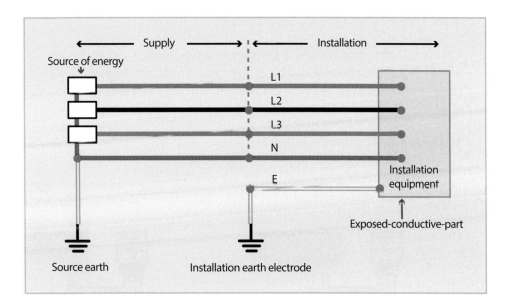

It is important to note that in a TT system the exposed-conductive-parts of the installation are connected by a protective conductor in the installation to the MET, and hence to the installation earth electrode, which is electrically independent of the source earth. These components of the installation provide a path for earth fault current to flow back to the source during a line-to-earth fault in the installation. Figure 4.11 shows the complete path that fault current will take under conditions of a line-to earth fault occurring in the installation

▼ **Figure 4.11** TT system showing an earth fault loop

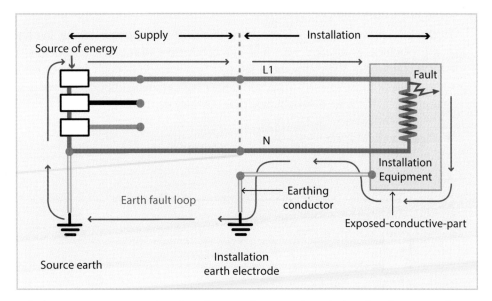

It is important to recognise that the supply to an installation forming part of a TT system may include a protective conductor. However, for an installation to form part of a TT system, the exposed-conductive-parts of the installation are required to be connected solely to the installation earth electrode.

The TT system is often used where a designer considers it undesirable to use the earthing facility offered by the electricity distributor, where it is from a PME supply network. Alternatively, in some cases, for statutory and/or for what are considered 'safety' reasons, the electricity distributor may be unwilling to offer an earthing facility. In this case, a TT system provides a solution.

It is worth noting that, as with a TN system, a supply to an installation forming part of a TT system can be provided by an overhead distribution system as well as, more commonly, by a supply cable buried underground. Figure 4.12 illustrates the two forms of supply.

▼ **Figure 4.12** TT system (single-phase) shown pictorially with an overhead and an underground supply cable, respectively

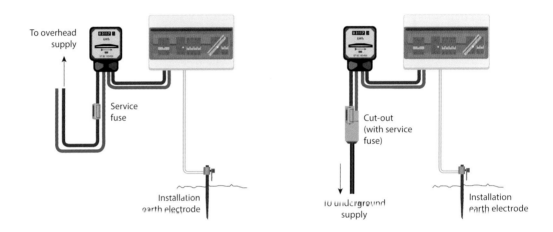

4.6 IT system

Figure 4.13 shows an IT system consisting of a single three-phase source and a three-phase electrical installation with four live conductors (three line conductors and a neutral conductor). The star point of the source is earthed through an earth electrode to the general mass of Earth via a sufficiently high earthing impedance. For a single-phase source, the source neutral would be similarly earthed.

▼ **Figure 4.13** IT system

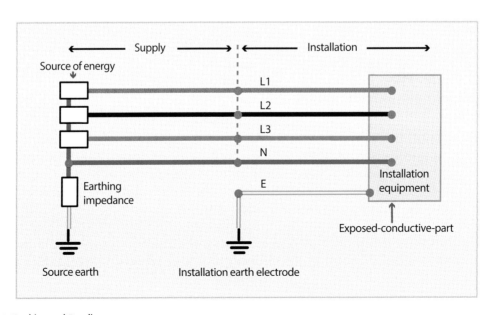

As the source of supply in an IT system is either isolated from Earth or earthed through a deliberately introduced high impedance, such a low voltage system is prohibited for use in public distribution networks in the UK by the ESQCR.

It is important to note that in an IT system the exposed-conductive-parts of the installation are connected by a protective conductor in the installation to the MET, and hence to the installation earth electrode, which is electrically independent of the source earth. These components of the installation provide a path for earth fault current to flow back to the source during a line-to-earth fault in the installation. Figure 4.14 shows the complete path that the fault current will take under conditions of a line-to-earth fault occurring in the installation.

▼ **Figure 4.14** IT system showing an earth fault loop

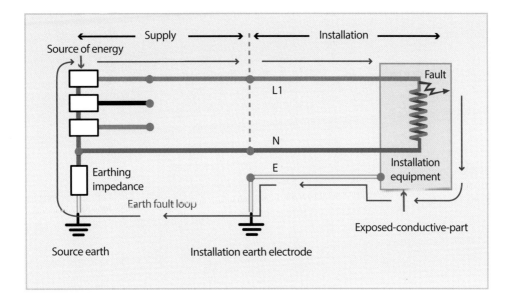

Protective equipotential bonding

5.1 The purpose of protective equipotential bonding

Protective equipotential bonding is not to be confused with earthing. Bonding serves the function of minimizing the magnitude of touch voltages within the building when an earth fault occurs in the installation. For a TN-C-S system, it will reduce touch voltages in the event of an open-circuit PEN fault on the supply.

Part 2 Touch voltages occur when an earth fault develops in the installation. *Earth fault current* is defined in BS 7671 as:

A current resulting from a fault of negligible impedance between a line conductor and an exposed-conductive-part or a protective conductor.

As this current flows to earth, touch voltages can be generated by the impedances between:

▶ two or more exposed-conductive-parts;
▶ two or more extraneous-conductive-parts;
▶ exposed-conductive-parts and extraneous-conductive-parts; and
▶ exposed-conductive-parts and Earth.

▼ **Figure 5.1** Example of touch voltages occurring in an installation under earth fault conditions

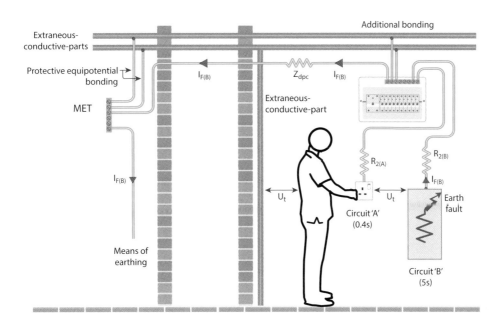

As Figure 5.1 illustrates, an earth fault on circuit B produces a touch voltage, U_t, between exposed-conductive-parts of that circuit and exposed-conductive-parts of circuit A and, in turn, to the extraneous-conductive-parts. Equation (5.1) can be used to predict the touch voltage, U_t, from the nominal voltage, U_0, the total earth fault loop impedance, Z_s, and the resistance of the circuit protective conductor in circuit B, i.e. $R_{2(B)}$:

$$U_t = \frac{U_0}{Z_s} R_{2(B)} \left(V\right)$$

(5.1)

Whereas the purpose of earthing is to limit the duration of the touch voltages, the purpose of protective equipotential bonding is to minimize the magnitude of these voltages. In essence, protective equipotential bonding will limit touch voltages to acceptable levels until the earth fault is automatically disconnected.

There are two protective measures associated with protective equipotential bonding:

▶ ADS (automatic disconnection of supply): this measure is used extensively in electrical installations in the UK and elsewhere. Figure 5.2 illustrates an installation where there are main protective bonding conductors and, in a location where there is an increased risk of electric shock, supplementary bonding conductors.
▶ earth-free local equipotential bonding: this is another protective measure providing fault protection. Unlike ADS, this measure is required to be under effective supervision. This seldom-used application is restricted to laboratories, electronic workshops and the like, where an earth-free environment is required.

Supplementary equipotential bonding, sometimes referred to as additional bonding, is also employed as an additional protective measure to ADS. It is required in certain of the special installations or locations addressed in Part 7 of BS 7671. It is also used where the conditions for automatic disconnection of supply are not feasible by means of an overcurrent protective device, as referred to in Regulation 411.3.2.5. It should, however, not be a designer's first choice of protection, as its use generally involves tolerating disconnection times longer than permitted by BS 7671, albeit with reduced touch voltages between the connected parts. In addition, Regulation 415.2 (note 2) states that 'such use of supplementary bonding does not exclude the need to disconnect the supply for other reasons, for example protection against fire, thermal stresses in equipment etc.'. Figure 5.2 illustrates examples of supplementary bonding conductors. See Chapter 8 for further information on this measure.

▼ **Figure 5.2** Examples of main and supplementary bonding conductors

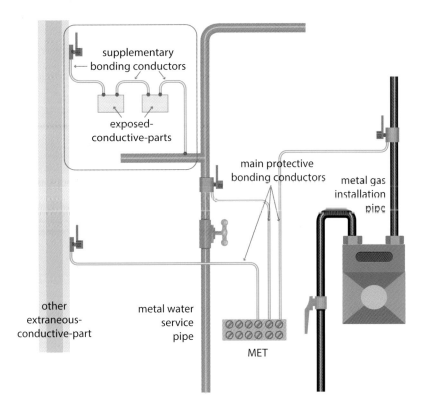

ADS is the most commonly used measure for protection against electric shock. A prerequisite for this measure to be effective is that the installation is required to be earthed, as well as having protective equipotential bonding. During an earth fault, earthing permits an earth fault current to flow and to be detected by a device provided to automatically disconnect the circuit (e.g. an overcurrent protective device or an RCD). The provision of earthing, together with a suitable protective device, limits the duration of an earth fault to acceptable limits. On the other hand, protective equipotential bonding minimizes the touch voltages that may appear between exposed-conductive-parts and/or extraneous-conductive-parts, thereby together preventing the occurrence of dangerous voltages.

411.1(ii) The protective measure of ADS encompasses three separately identifiable components
411.3 to provide fault protection:

▶ earthing of installation equipment metalwork (exposed-conductive-parts);
▶ automatic disconnection of supply; and
▶ protective equipotential bonding, as required, of non-electrical metalwork (extraneous-conductive-parts).

Figure 5.3 illustrates part of an electrical installation (the type of system of which it forms a part is not important here). In the circuit shown, an earth fault has developed in the current-using equipment. As a result, a fault current (I_f) flows along the circuit protective conductor and back to the source. A small proportion of the current may flow through the main protective bonding conductor directly to Earth and back to the source.

▼ **Figure 5.3** Illustration of how protective bonding reduces touch voltages under fault conditions

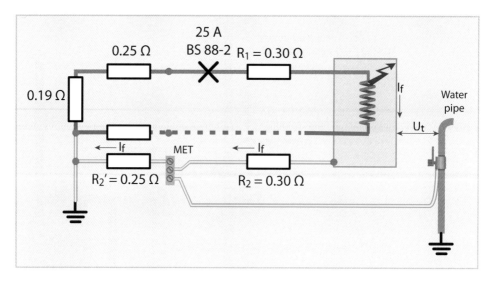

The touch voltage (U_t) with protective equipotential bonding, or the potential difference, between the exposed-conductive-parts of the equipment and the simultaneously accessible extraneous-conductive-part is given by Equation (5.2):

$$U_t = I_f \times R_2 = \frac{U_0}{7_s} \times R_2 (V)$$
(5.2)

The equation ignores the reactance of the circuit protective conductor, any small effect due to current flowing in the main protective bonding conductor, and the voltage factor (C_{max} or C_{min}).

To apply Equation (5.2), take the example of a circuit that supplies an item of current-using equipment rated at 230 V (U_0) and 5 kW (resistive load). The circuit is protected by a 25 A, BS 88-2 fuse and the earth fault loop impedance Z_s is 1.29 Ω (the limiting value from Table 41.2 of BS 7671). R_1 is 0.30 Ω and R_2 is also 0.30 Ω. The installation forms part of a TN-S system where the external earth fault loop impedance, Z_e, is 0.69 Ω, this being made up of 0.25 Ω for the supply protective conductor, 0.19 Ω for the source of energy such as the transformer winding and 0.25 Ω for the supply line conductor.

The calculated U_t without protective equipotential bonding, is given in (5.3):

$$U_t = I_f \times (R_2 + R_2') = \frac{U_0}{Z_s} \times (R_2 + R_2') = \frac{230}{1.29} \times (0.30 + 0.25) = 98.1V$$
(5.3)

By connecting the MET to the extraneous-conductive-parts, U_t is minimized. Without this conductor, the potential difference would approximate to the voltage drop produced by the earth fault current, I_f, along the full length of the earth return path. This could be significantly greater than $I_f R_2$ because of the voltage dropped across the supply protective conductor (R_2'). Failure to provide all necessary main protective bonding conductors within an installation would almost certainly increase the risk of electric shock under fault conditions.

The calculated touch voltage with protective equipotential bonding is given in (5.4).

$$U_t = I_f \times R_2 = \frac{U_0}{Z_s} \times R_2 = \frac{230}{1.29} \times 0.30 = 53.5\,V \tag{5.4}$$

In the case cited in Equation (5.3), and assuming that the impedance of the supply protective conductor is 0.25 Ω, the voltage dropped along this conductor would be approximately 44.6 V (0.25 × U_0/Z_s). Had the protective equipotential bonding not been present, the touch voltage would have been approximately 98 V. With the bonding present, the voltage is 53 V.

The ADS protective measure is achieved by coordinating the characteristics of the protective device for automatic disconnection and the relevant impedance of the circuit concerned, but the requirements for main protective bonding conductors and, where applicable, supplementary equipotential bonding, are required to be met. Automatic disconnection, a necessary constituent of ADS, is addressed in Chapter 7 of this Guidance Note.

5.2 Main protective bonding conductors

Main bonding or, more correctly, main protective equipotential bonding, is required in most electrical installations. Such bonding is an essential part of the most commonly used measure of protection against electric shock, ADS.

411.3.1.2
Chap 54 The first part of Regulation 411.3.1.2 requires that in each installation main protective bonding conductors complying with Chapter 54 shall connect to the main earthing terminal extraneous conductive-parts including the following:

▶ water installation pipes;
▶ gas installation pipes;
▶ other installation pipework and ducting;
▶ central heating and air conditioning systems; and
▶ exposed metallic structural parts of the building.

With the introduction of BS 7671:2018, a new requirement has been added pointing out that metallic pipes entering the building having an insulating section at their point of entry need not be connected to the protective equipotential bonding.

The bonding of a lightning protection system shall be carried out in accordance with BS EN 62305 (see 5.4).

Regulation 411.3.1.2 goes on to require that 'where an installation serves more than one building the above requirement shall be applied to each building.'

The final paragraph of Regulation 411.3.1.2 stipulates that 'to comply with these Regulations it is also necessary to apply equipotential bonding to any metallic sheath of a telecommunication cable. However, the consent of the owner or operator of the cable shall be obtained.'

Part 2 This regulation requires all extraneous-conductive-parts to be bonded to the MET. An extraneous-conductive-part is defined in Part 2 of BS 7671 as:

A conductive part liable to introduce a potential, generally Earth potential, and not forming part of the electrical installation.

Chapter 6 addresses this definition in more detail.

It should be noted that, whilst a number of examples of extraneous-conductive-parts are given in the regulation, the list is not exhaustive and other items not listed may well fall within the definition.

Extraneous-conductive-parts may be, and often are, connected to the MET individually, as shown in Figure 5.4. However, it is permitted to connect them collectively or in groups where the main protective bonding conductor is looped from one extraneous-conductive-part to another. Where bonding is undertaken in this way, the bonding conductor should remain unbroken at intermediate points, as shown in Figure 5.5, thus maintaining continuity to other extraneous-conductive-parts should one be disconnected for any reason.

▼ **Figure 5.4** The MET

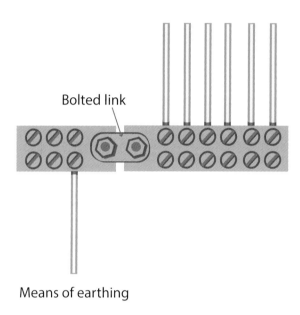

▼ **Figure 5.5** An unbroken main protective bonding conductor

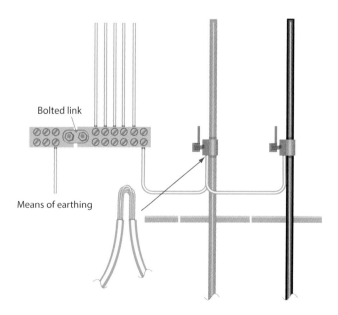

Figure 5.6 shows typical main protective bonding conductors connecting extraneous-conductive-parts separately to the MET and hence to the means of earthing.

▼ **Figure 5.6** A typical example of protective equipotential bonding

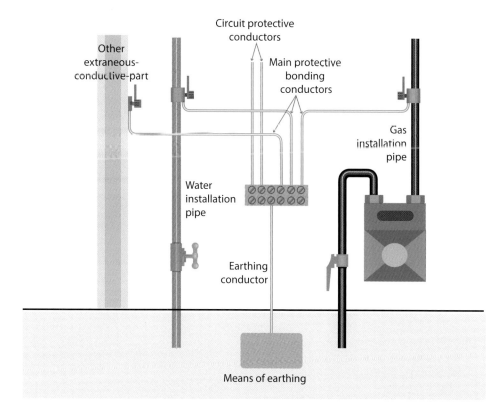

There are a number of types of conductor that are suitable for use as a main protective bonding conductor, although a single-core non-flexible copper cable with a green-and-yellow covering is probably the most commonly used.

543.2.1
543.2.3
Regulation 543.2.1 also permits the use of other types of conductor, including metal parts of wiring systems such as metal conduits and metal trunking, as well as metal sheaths or armouring of cables, provided that all the relevant requirements of BS 7671 are met. However, neither a metallic gas nor an oil pipe may be used for such a purpose (see Regulation 543.2.3). Similarly, flexible or pliable conduit is precluded for such use by this regulation.

For installations forming part of a TN-C-S system in which the supply is earthed at multiple points (PME conditions applying), where it is intended to utilize the armouring of a cable or a core of an armoured cable as a main protective bonding conductor, the designer needs to consider the effects of the currents that may flow in the conductor due to network conditions. The armouring or a core of such a cable is normally not used for this purpose where PME conditions apply, unless the electrical installation designer determines that the heat produced in the armouring or core due to its use will not cause overheating of the live conductors of the cable when on full load.

5.2.1 Cross-sectional area

544.1.1
542.3.1
Regulation 544.1.1 sets out the requirements for the minimum permitted csa of a main protective bonding conductor. Where such a conductor is buried in the ground, the additional requirements of Regulation 542.3.1 for earthing conductors should be applied.

5

544.1.1 For installations where PME conditions do not apply, Regulation 544.1.1 requires main protective bonding conductors to have a csa of:

▶ not less than half that required for the earthing conductor; and
▶ not less than 6 mm², but need not be more than 25 mm² if the conductor is of copper or, if of another metal, a csa affording equivalent conductance.

Table 54.7 This regulation stipulates that the csa of the main protective bonding conductors should be related to the csa of the earthing conductor. Where the csa of the earthing conductor is selected using Table 54.7 of BS 7671, as is normally the case, then Table 5.1 provides data relating the size of the main protective bonding conductors to the supply line conductor.

▼ **Table 5.1** Csa of earthing conductors and main protective bonding conductors for installations where PME conditions do not apply

Line conductor[1] csa (mm²)	Earthing conductor[1,2] csa (mm²)	Main protective bonding conductor[1] csa (mm²)	Comment
4	4	6	Main protective bonding conductor csa must have a minimum value of 6 mm²
6	6	6	
10	10	6	
16	16	10	
25	16	10	
35	16	10	
50	25	16	
70	35	25	Main protective bonding conductor csa need not be more than 25 mm²
95	50	25	
120	70	25	
150	95	25	
185	95	25	
240	120	25	
300	150	25	
400	240	25	

1 Assumes that the conductors are copper.
2 Other constraints may apply to the csa of the earthing conductor (see Chapter 3).

For a non-copper main protective bonding conductor, the 'equivalent conductance' requirement will be met if the csa (S_m) of the conductor is not less than that given by Equation (5.5):

$$S_m \geq S_c \left(\frac{\rho_m}{\rho_c} \right) (mm^2) \tag{5.5}$$

where:

S_m is the minimum csa required for the main protective bonding conductor (in a metal other than copper).

S_c is the minimum csa required for a copper main protective bonding conductor.

ρ_m is the resistivity of the metal from which the main bonding conductor is made.

ρ_c is the resistivity of copper.

For a steel main protective bonding conductor, the ratio ρ_m/ρ_c is 8.0. Similarly, for an aluminium conductor, the ratio is 1.62. For conductors made of these metals, the csa would be required to be 8.0 and 1.62 times that of copper, respectively, in order to afford an equivalent conductance.

Where PME conditions apply

In the event of an open circuit PEN conductor of a TN-C-S system with PME, load currents may flow through the main protective bonding conductors and earthing of an installation.

Table 54.8
544.1.1 Consequently, for an installation where PME conditions apply, Regulation 544.1.1 requires the main protective bonding conductors to be selected in relation to the PEN conductor of the supply and Table 54.8. Regulation 542.3.1 also requires (amongst other things) that the earthing conductor shall meet those same requirements of Regulation 544.1.1. The data from Table 54.8 is available in Table 3.2 of this Guidance Note.

As the earthing conductor also performs the function of a main protective bonding conductor, the requirements of Regulation 544.1.1 should also be met for the earthing conductor, so that the csa of the earthing conductor is not less than that required for main protective bonding conductors, as well as meeting the requirements of Regulation 542.3.1.

The electricity distributor may have particular requirements for the csa of main protective bonding conductors, which may exceed the minimum csas given in Table 54.8. If there is doubt in this respect, the electricity distributor should be consulted and guidance sought at an early stage of the design.

The PEN conductor referred to in column 1 of Table 3.2 is that of the electricity distributor's low voltage supply to the installation. This is the combined protective and neutral PEN or CNE conductor of the supply. It is not the neutral conductor on the consumer's side of the supply terminals, which may have a smaller csa, as might be the case in a downstream distribution circuit feeding a separate building.

Where the use of non copper main protective bonding conductors is contemplated, the advice of the electricity distributor should be sought and followed.

5.2.2 Identification

514.3.1
514.4.2 For a main protective bonding conductor consisting of a single-core cable or a core of a cable, Regulations 514.3.1 and 514.4.2 require it to be identified with the bi-colour combination green-and-yellow, as shown in Figure 5.8. A bare conductor, such as a tape, strip or bare stranded conductor, is required by Regulation 514.4.2 to be identified where necessary at intervals with the green-and-yellow colour combination.

514.1.2 As with all wiring, main protective bonding conductors have to be identifiable for inspection, testing, repair or alteration of the installation (see Regulation 514.1.2). To fulfil this requirement, each bonding conductor is required to be marked or labelled to indicate its function, and to identify the item(s) to which it connects (e.g. a structural steel or water service pipe), unless this is clear from the arrangement of the conductor, as might be the case from its position or from the terminals to which it connects, as illustrated in Figure 5.7.

▼ **Figure 5.7** Identification of protective conductors by the green-and-yellow colour scheme and labels

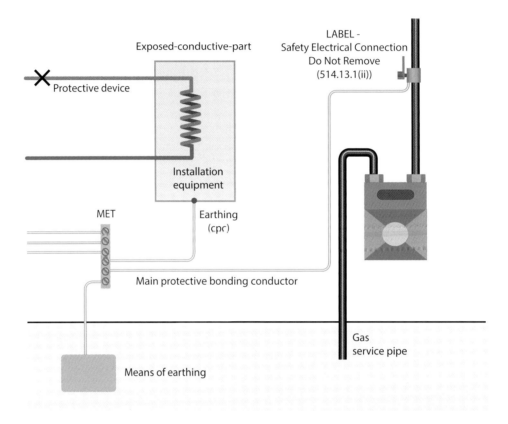

5.2.3 Supports

Where a main protective bonding conductor forms part of a composite cable, such as a separate core or the metal armouring or sheath, the method of support for the bonding conductor will be dictated by the type of cable and the manufacturer's installation instructions.

543.3.1 However, single-core bonding conductors, and similar tape conductors, are required to be adequately supported, avoiding non-electrical services such as pipework as a means of support, so that they are able to withstand mechanical damage etc. (see Section 522 of BS 7671) and any anticipated factors likely to result in deterioration (see Regulation 543.3.1).

Table 5.2 gives guidance on providing adequate support for both main protective bonding conductors and supplementary bonding conductors. Figure 5.8 illustrates a main protective bonding conductor supported in an acceptable manner.

▼ **Table 5.2** Recommended spacing of supports of single-core rigid copper cables for protective bonding conductors

Overall cable diameter, d (mm)	Horizontal spacing (mm)	Vertical spacing (mm)	Comment
d ≤ 9	250	400	This data is provided as a guide and may be overridden by the constraints of good workmanship and by visual considerations
9 < d ≤ 15	300	400	
15 < d ≤ 20	350	450	
20 < d ≤ 40	400	550	

▼ **Figure 5.8** Example of support for main protective bonding conductor

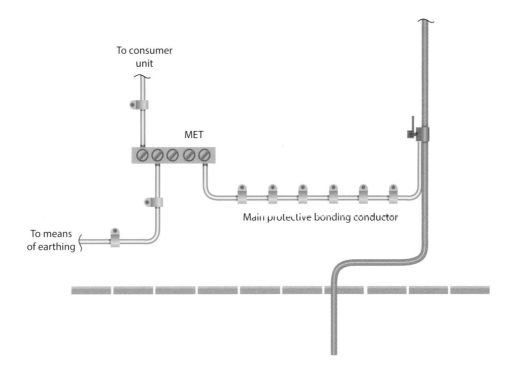

To consumer unit

MET

To means of earthing

Main protective bonding conductor

5.2.4 Additions and alterations

Where additions and/or alterations to an existing installation that pre-dates the current requirements of BS 7671 with regard to protective equipotential bonding are to be performed, it is important to fully consider any shortcomings. For example, it may be that not all extraneous-conductive-parts are bonded or that the csa of the conductors is insufficient to meet the current requirements.

Where there is an absence of main protective bonding conductors to certain extraneous-conductive-parts, the right course of action is to install suitable bonding conductors. Where the deficiency is that the csa of the conductor is less than currently required, the designer needs to consider carefully whether retention of the existing bonding conductors will be able to provide adequate safety levels and, if not, whether the conductors should be upgraded to meet current requirements.

132.16 Where modifications to an existing installation are planned, the starting point should be the requirement of Regulation 132.16, reproduced here for convenience:

No addition or alteration, temporary or permanent, shall be made to an existing

installation, unless it has been ascertained that the rating and the condition of any existing equipment, including that of the distributor, will be adequate for the altered circumstances. Furthermore, the earthing and bonding arrangements, if necessary for the protective measure applied for the safety of the addition or alteration, shall be adequate.

It is clear that it is essential for the protective equipotential bonding to be adequate. Where such bonding is not present, it should be provided. It would be wholly unacceptable to omit main protective bonding conductors or to rely on bonding conductors with an inadequate csa.

In the case where all main protective bonding conductors are present, but their csa is less than currently required, the installation designer should make an evaluation of their suitability for use under the proposed new conditions.

120.3   Having considered the adequacy of the existing bonding conductors, an installation designer may, in those cases in which the resulting degree of safety is considered to be not less than that obtained by compliance with BS 7671, elect to retain the existing conductors. However, the departure is required to be recorded on the certification for the work, both in the 'Details of departures' and in the 'Comments on the existing installation' (see Regulation 120.3).

Table 54.8   For an installation forming part of a TN-C-S system with PME, a careful assessment should be made before relying on bonding conductors with an inadequate csa. For such installations, Table 54.8 of BS 7671 sets out the minimum csa, based on those given in the Electricity Supply Regulations 1988 (superseded by the ESQCR). Prior to the publication of these statutory regulations, smaller minimum csas were accepted by some electricity boards. However, the minimum csas given in Table 54.8 of BS 7671 were intended to remove the likelihood of protective bonding conductors overheating due to diverted load currents in the event of an open circuit PEN conductor in the supply. Only where the designer is comfortable with the adequacy of the existing bonding conductors should the decision be made not to replace them with ones meeting the current minimum requirements.

Consideration should also be given to the requirements of the distribution network operator (DNO).

418.2   ## 5.3   Earth-free local equipotential bonding

Earth-free local equipotential bonding is one of the protective measures for fault protection permitted by BS 7671.

This protective measure is intended to prevent the appearance of a dangerous touch voltage between simultaneously accessible conductive parts in the event of a failure of the basic insulation. Such a protective measure must only be installed where the installation is controlled or under the supervision of an electrically skilled or instructed person or persons, to prevent unauthorized changes being made.

418.2.5   Local protective equipotential bonding is required to be carried out by connecting together all the exposed-conductive-parts and extraneous-conductive-parts within the location of the installation. It is important that this bonding is not connected to Earth. In this way, persons within the location are free from electric shock. However, where a person enters or leaves the equipotential location, there may be a risk of electric shock; BS 7671 requires precautions to be taken to address this risk. This includes a requirement to place a notice in a prominent position adjacent to every point of access to the location, as illustrated in Figure 5.9.

This measure is strictly limited to situations that need to be earth-free. Examples include medical, special electronic and communications equipment applications, in which the equipment, while having exposed-conductive-parts (i.e. not being Class II equipment or having equivalent insulation), will not operate satisfactorily where connected to a means of earthing. The locations where the measure is applied may have a non-conducting floor, or a conductive floor which is insulated from earth and to which every accessible exposed-conductive-part within the location is connected by local protective bonding conductors. The location requires effective supervision and regular inspection and testing of the protective features.

The correct implementation of the requirements results in a 'Faraday cage' and prevents the appearance of any dangerous voltages between simultaneously accessible parts within the location concerned. Thus, in this application of equipotential bonding, the use of the word 'equipotential' really does reflect what it means for a first fault on the equipment, although this may not hold true for two simultaneous faults.

This measure cannot realistically be applied to an entire building and it is difficult to co-ordinate safely with other protective measures used elsewhere in the installation. In particular, precautions are necessary at the threshold of the earth-free equipotential location. In addition, as mentioned earlier, a warning notice is required to be fixed at every point of access into the location to warn against the importation of an earth.

The form of the supply to the equipment requires special consideration. Using the mains supply of a TN system would import an earth into the location via the earthed neutral conductor which, in the event of a neutral fault to an exposed-conductive-part, would earth the equipment.

Earth-free local equipotential bonding is normally associated with electrical separation, which overcomes this problem. However, where two measures of protection are to be used in the same location, care should be taken to ensure that the particular requirements for each measure are fully satisfied and, most important, are mutually compatible.

418.2
514.13.2   Not surprisingly, this protective measure is not suitable for locations where there is perceived to be an increased risk of electric shock, such as those addressed in Part 7 of BS 7671, refer to regulations 418.2 and 514.13.2.

Regulations 418.2 and 514.13.2 refer.

▼ **Figure 5.9** Label for earth-free location

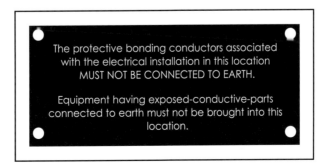

The protective bonding conductors associated with the electrical installation in this location MUST NOT BE CONNECTED TO EARTH.

Equipment having exposed-conductive-parts connected to earth must not be brought into this location.

5.4 Bonding of lightning protection systems

411.3.1.2 Although BS 7671 excludes lightning protection systems of buildings from its scope, there is a requirement in Regulation 411.3.1.2 to provide a protective bonding connection of such systems to the MET of the electrical installation, in accordance with BS EN 62305. The requirements in terms of csa of this bonding are as for all other extraneous-conductive-parts. The recommendations of BS EN 62305 are also required to be taken into account.

Based on the guidance given in BS EN 62305, the positioning of protective bonding connections to a lightning protection system is important and best determined by a lightning protection system designer.

Where the down conductors of the lightning protection system are systematically connected together, a single protective bonding connection to that system will suffice. A connection from the MET would normally be taken to the closest down conductor by the most direct route.

Although each bonding connection arrangement should be considered on its own merits in consultation with all the authorities concerned, Figure 5.10 shows a typical arrangement.

▼ **Figure 5.10** Typical arrangement for the main protective bonding conductor connecting to a lightning protection system

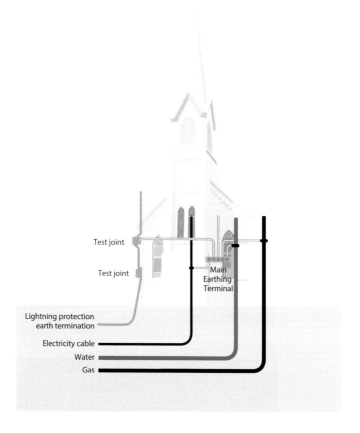

Additionally, it may be necessary to connect parts of the electrical installation to adjacent parts of the lightning protection system which are separated by less than the predicted isolation distance necessary to prevent side-flashing.

Bonding connections to a lightning protection system are often made outdoors. Where this is the case, attention should be paid to any external influences that might affect the connection. This is particularly important where conductors with different shapes and made from different metals are used. Precautions may need to be taken to avoid corrosion and to create joints with an adequate mechanical strength and durable electrical continuity.

5.5 Extraneous-conductive-parts common to a number of buildings

Industrial and commercial facilities sometimes consist of a number of separate buildings supplied from a common electricity source, or separately fed with their own supplies. It is often a feature of these premises to have extraneous-conductive-parts that are common to a number of separate buildings. For example, a common metal pipework system transporting gases, liquids or steam between buildings would require careful consideration. Figure 5.11 illustrates three buildings that have a common pipework system.

▼ **Figure 5.11** Three buildings with a common pipework system

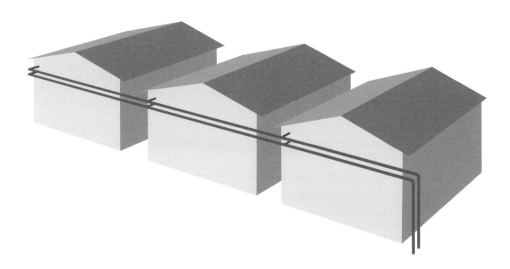

In these circumstances, the question of protective equipotential bonding of the extraneous-conductive-parts needs careful thought. The guidance given here relates to installations fed from a single low voltage source (a single low voltage distribution transformer). Two situations are considered:

▶ Situation 1: the three buildings are supplied from a common source, with the electricity distributor's supply to one of the buildings and distribution circuits emanating to the other two buildings (Figure 5.12).
▶ Situation 2: the three buildings are supplied from a common source, with the electricity distributor's separate electricity supply to each of the three buildings (Figure 5.13).

▼ **Figure 5.12** Three buildings with a single electricity supply

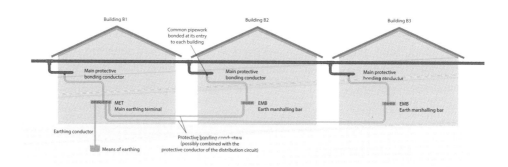

▼ **Figure 5.13** Three buildings with three separate electricity supplies

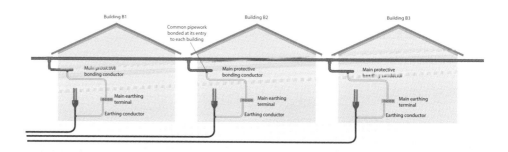

Where multiple sources, including sources from local transformers, are involved, further advice should be sought from a suitably qualified electrical engineer.

A metal pipework system that entered a building would be considered to be an extraneous-conductive-part. This would be so even where the pipework had previously exited another building in which it had been separately bonded elsewhere. The pipework would then be likely to introduce a potential, either Earth potential or earth fault potential, from the building into which the pipework previously entered. Hence it would fall within the definition of an extraneous-conductive-part.

There would therefore be a requirement to provide protective equipotential bonding to the common pipework system in each building that it entered, either to the MET if the origin of the supply was there, or to the earth marshalling bar (EMB), if in another building.

411.3.1.2 With the introduction of BS 7671:2018, a new requirement has been added pointing out that metallic pipes entering the building having an insulating section at their point of entry need not be connected to the protective equipotential bonding.

In Figures 5.12 and 5.13, each building will have its own earth marshalling point, either in the form of an MET or an EMB, with the common pipework system protective equipotentially bonded to the MET or EMB in each building:

In situation 1, the MET is in building B1 and there are EMBs in buildings B2 and B3. The common pipework system is main bonded:

▶ to the MET in building B1; and
▶ to the EMBs in buildings B2 and B3.

In situation 2, an MET exists in buildings B1, B2 and B3. The common pipework system is main bonded:

▶ to the MET in building B1;
▶ to the MET in building B2; and
▶ to the MET in building B3.

544.1.2 The position of the connection of the main protective bonding conductors to the common pipework system should be as close as practicable to the point of entry of the pipework into each building.

The protective bonding conductors between the MET and the two EMBs may be made by utilizing the circuit protective conductors of the distribution circuits to buildings B2 and B3, provided that the circuit protective conductors meet the csa requirements for both functions.

5.6   Installations serving more than one building

411.3.1.2 Where a supply serves installations in more than one building, Regulation 411.3.1.2 requires that main protective bonding conductors should be applied in each building. The extraneous-conductive-parts in all of the buildings should therefore be connected to the MET, of which there is only one.

Sect 544 So, where a supply serves three separate buildings, designated B1, B2 and B3, each will be required to be provided with main protective bonding conductors to all extraneous-conductive-parts, complying with Section 544 of BS 7671.

In such a scenario, an MET is located in building B1 at a position adjacent to the incoming supply at the origin of the installation. The two other buildings each have EMBs. An EMB marshals all the main protective bonding conductors for a particular building and also provides for the connection of a conductor that runs to the MET.

As with any protective equipotential bonding, the csa of the conductors will depend on the system type. Figure 5.14 illustrates a three-building complex where the system type is TN-S for buildings B1 and B3, and TT for building B2. Figure 5.15 illustrates a three-building complex where the system type is TN-C-S with PME for buildings B1 and B3, and TT for building B2.

544.1.1 For buildings B1 and B3, it is important to note that, irrespective of the system type, the csa of the main protective bonding conductors should be not less than that given in Regulation 544.1.1. This applies equally to the conductors in both buildings in the examples given in Figures 5.14 and 5.15. For the three system types cited, the requirements in terms of csa of the main protective bonding conductors are:

▶ for TN-S and TT systems: the csa of the bonding conductors is required to be not less than half the csa required for the earthing conductor, subject to a minimum of 6 mm². It need not exceed 25 mm² if the conductor is made of copper (or affording an equivalent conductance in other metals). Note that the earthing conductor, of which there is but one, is the conductor that connects the MET to the means of earthing, which in this case is the electricity distributor's earthing terminal.

Table 54.8 ▶ for TN-C-S systems: where PME conditions apply, the csa of the bonding conductors has to be selected in relation to the supply neutral conductor and in accordance with Table 54.8 of BS 7671. Note that the supply neutral conductor is the neutral conductor upstream of the electricity distributor's cut-out.

Both figures illustrate three separate buildings, each with a single extraneous-conductive-part in the form of an incoming metal service pipe. In reality, other extraneous-conductive-parts, such as structural steel and/or lightning protection systems, may be present, but for simplicity and clarity in the figures, only one such item is shown.

544.1.1 The application of Regulation 544.1.1 for complexes with a number of separate buildings can result in the csa of the main protective bonding conductors being required to be greater than the csa of the associated live conductors of the distribution circuit supplying that building, particularly where PME conditions apply. This is illustrated in Figure 5.15, where the main protective bonding conductor in building B3 is required to have a csa of 35 mm², as the csa of the supply neutral is 120 mm², which is considerably greater than the required csa of 16 mm² for the circuit protective conductor of the distribution circuit to building B3.

543.1
544.1.1 Where a circuit protective conductor also acts as a bonding conductor, the requirements for both functions will have to be met. In other words, as well as meeting the requirements of Regulation 543.1 for a circuit protective conductor, the requirements of Regulation 544.1.1 must be met. This protective conductor dual function has been adopted for the distribution circuit to building B3 in Figures 5.14 and 5.15.

542.1.3.3 For the case of building B2, where the distribution circuit includes a circuit protective conductor, as might be the case with an armoured cable, it is important to consider Regulation 542.1.3.3, which reads:

Where a number of installations have separate earthing arrangements, any protective conductors common to any of these installations shall either be capable of carrying the maximum fault current likely to flow through them or be earthed within one installation only and insulated from the earthing arrangements of any other installation. In the latter circumstances, if the protective conductor forms part of a cable, the protective conductor shall be earthed only in the installation containing the associated protective device.

411.3.1.1 In the examples shown in Figures 5.14 and 5.15, the distribution circuit protective conductor (cable armouring) has been insulated from the earthing arrangement of building B2 by means of an insulated cable gland. This effectively provides for the two earthing arrangements to be considered simultaneously inaccessible, thereby removing the need to apply Regulation 411.3.1.1, which requires the circuit protective conductor and the earthing arrangement to be connected together where they are simultaneously accessible, in order to prevent the associated risk of an electric shock due to the potential difference between the two points. However, the armouring would be required to be earthed in building B1.

542.1.3.3 ▼ **Figure 5.14** A three-building complex: system type TN-S
543.1.1
544.1.1
Table 54.1
Table 54.7

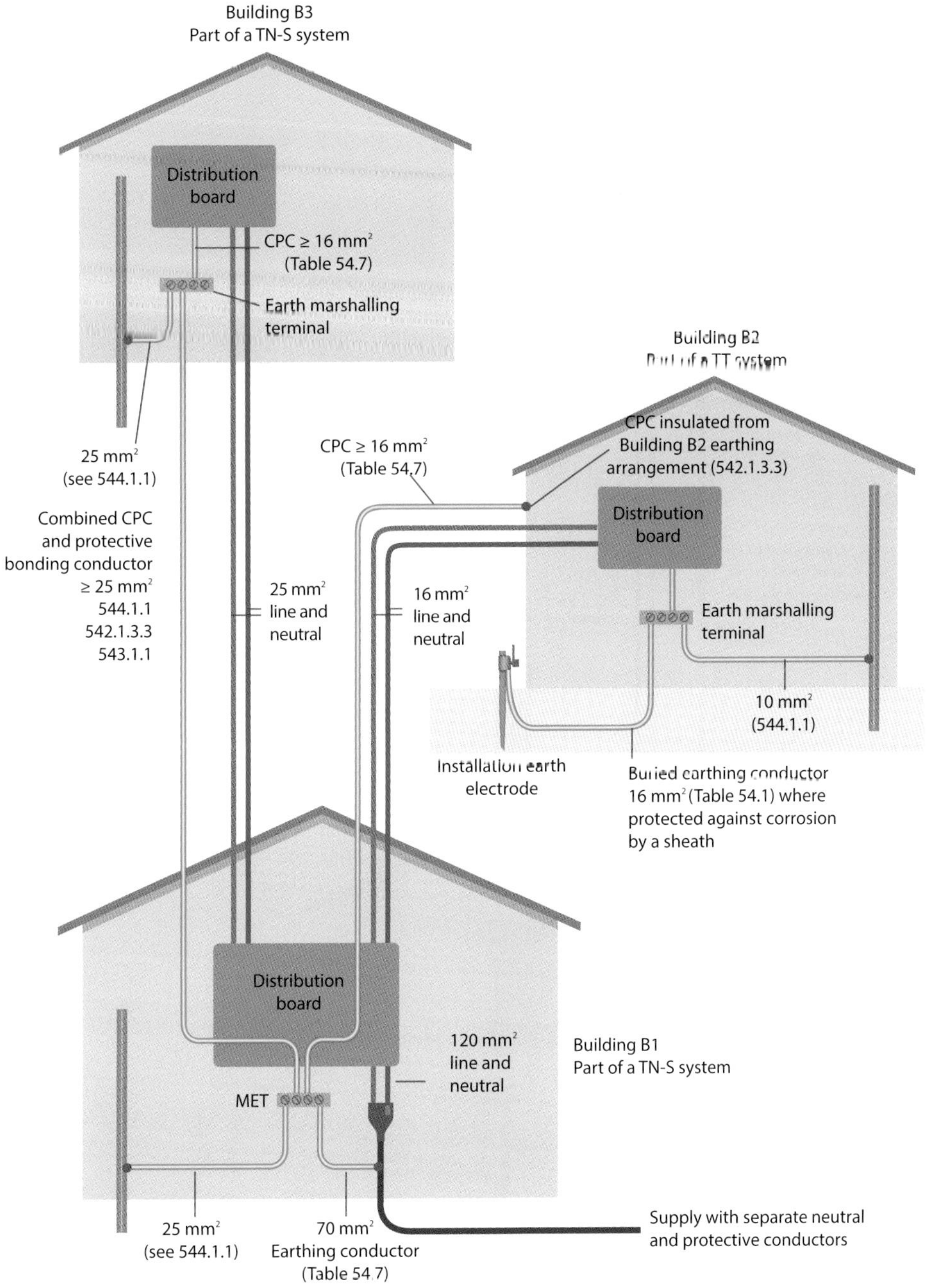

542.1.3.3
543.1.1
544.1.1
Table 54.1
Table 54.7

▼ **Figure 5.15** A three-building complex: system type TN-C-S

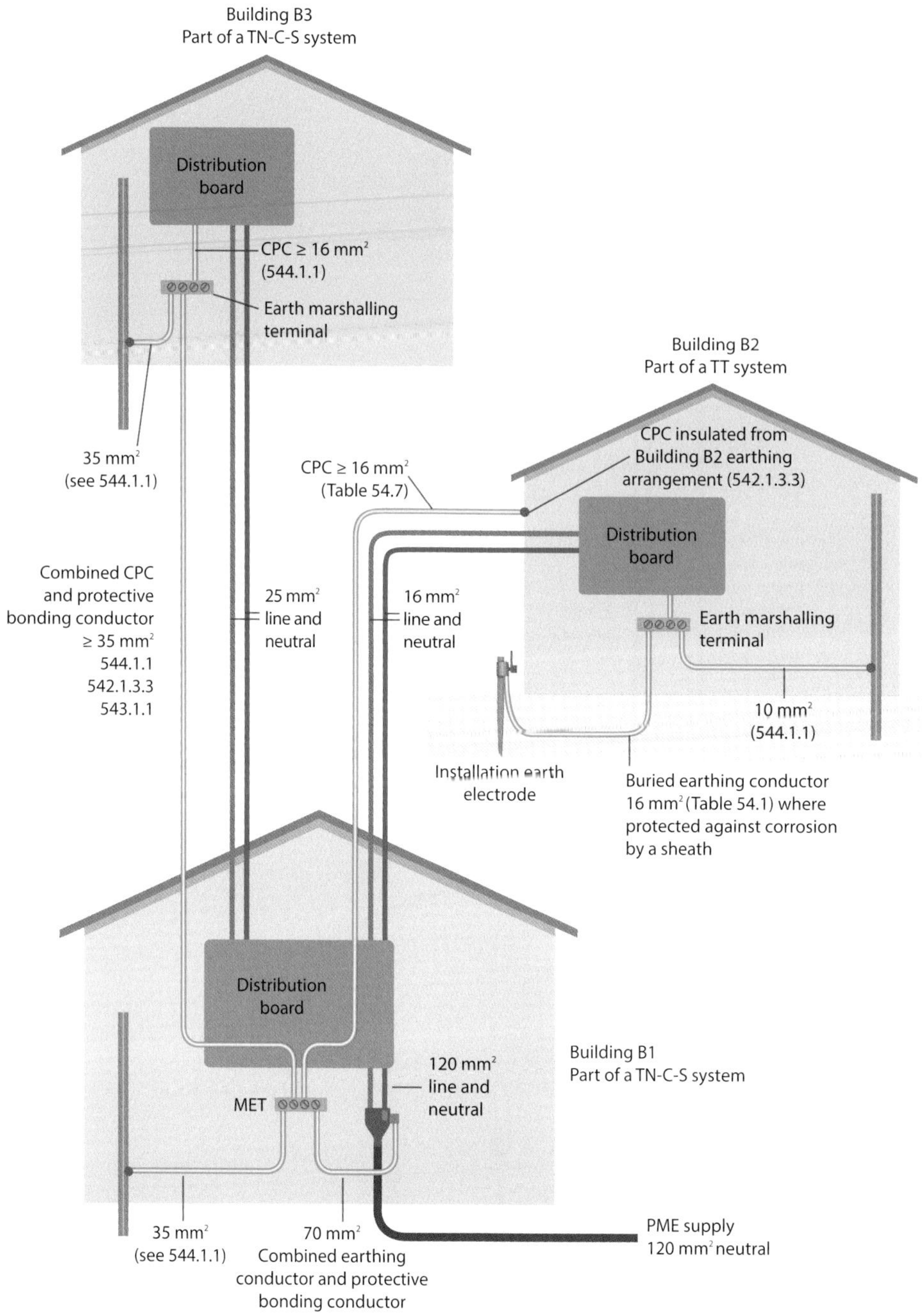

542.1.3.3 Where a protective conductor connection is to be provided between the earthing arrangement of building B2 and that of buildings B1 and B3, the protective conductor common to the two installations would be required by Regulation 542.1.3.3 to be capable of carrying the maximum fault current likely to flow through it. As a consequence, the magnitude and duration of the fault current would have to be calculated and, by using Equation (3.1) in section 3.2, the minimum csa required for the conductor would have to be determined.

For an installation fed from a supply to which PME conditions apply, but which forms part of a TT system as in building B2, these conditions do not apply, provided that the PME earthing facility has not been used to earth the installation.

5.7   Multi-occupancy premises

Protective equipotential bonding in multi-occupancy premises can sometimes present problems in terms of interpreting the requirements of BS 7671. This typically involves premises such as blocks of flats, office blocks and shops that are separately let and have separate electricity supplies. Additionally, single-occupancy properties that have subsequently been converted into separate units sometimes present problems too.

411.3.1.2 BS 7671 requires main protective bonding conductors in each installation. The definition of installation, reproduced below together with that for the origin of the installation, relates to an installation supplied from a common origin.

Part 2   **Electrical installation** (abbr: Installation). An assembly of associated electrical equipment having co-ordinated characteristics to fulfil specific purposes.

Part 2   **Origin of an installation**. The position at which electrical energy is delivered to an electrical installation.

It is clear that BS 7671 requires protective equipotential bonding in each and every installation connecting together extraneous-conductive-parts to the MET, and this would apply equally to separate installations of a multi-occupancy building.

544.1.1 The csa of every main protective bonding conductor has to be in accordance with Regulation 544.1.1 and will depend on the system type as discussed in clause 5.2.1. As previously mentioned, except where PME conditions apply, the csa of the protective bonding conductors is related to the csa of the earthing conductor of the installation (i.e. in each separate installation). For installations to which PME conditions apply, the csa of the protective bonding conductors is related to the electricity distributor's supply neutral conductor (not the neutral conductor downstream of the cut-out on the consumer's side, which may have a different csa).

Figure 5.16 illustrates a typical protective equipotential bonding arrangement in a small office block. Each office installation forms part of a TN-C-S system with a separate supply with PME applied.

As with all electrical installations, the electricity distributor may have additional requirements for protective equipotential bonding, especially in the case where PME conditions apply. It is therefore important that the installation designer seeks guidance in this respect from the electricity distributor.

544.1.2 The connection of main protective bonding conductors to extraneous-conductive-parts, such as gas and water service pipes, should be made in accessible locations as near as practicable to the point of entry into the office building, as specified by Regulation 544.1.2.

▼ **Figure 5.16** Typical protective equipotential bonding arrangement in a small office block

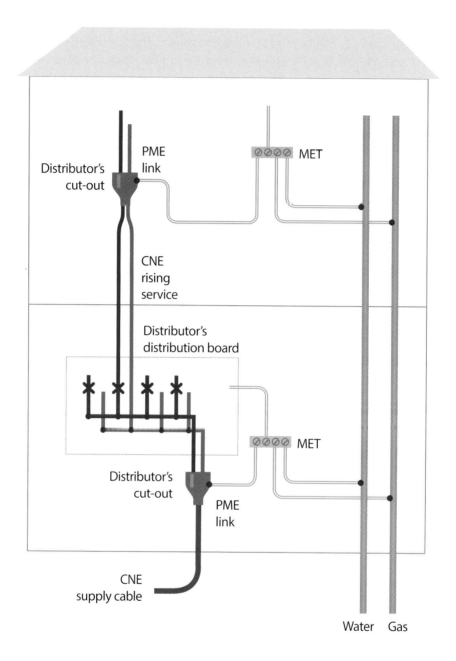

Extraneous-conductive-parts and their connections

6

6.1 Definition of an extraneous-conductive-part

It is important to note that the term 'extraneous-conductive-part' is hyphenated: it is therefore a single term. It has a special meaning which could not be attributed had it not been so hyphenated.

The term is used extensively with regard to protective equipotential bonding and automatic disconnection, and is also mentioned in connection with other measures for protection against electric shock, such as separated extra-low voltage (SELV), placing out of reach and non-conducting locations.

Part 2 An *extraneous-conductive-part* is defined in Part 2 of BS 7671 as:

> *A conductive part liable to introduce a potential, generally Earth potential, and not forming part of the electrical installation.*

Even with such a definition, it is sometimes difficult to establish what is and what is not an extraneous-conductive-part. To assist in making this decision, the installation designer has to analyse the definition by breaking it down into three discrete parts:

▶ a conductive part;
▶ liable to introduce a potential, generally Earth potential; and
▶ not forming part of the electrical installation.

Generally, it is metals with their high conductivity values that are described as the conductive part in the context of considering an extraneous-conductive-part. The conductivity of a material is a measure of its ability to conduct electricity or, put another way, of its conductance, G, which is the reciprocal of resistance (as conductivity is the reciprocal of resistivity).

However, although non-metals are generally non-conductive, pipework systems that are intended to carry fluids may not be so described. Water in its purest form is a good insulating material, with a 10^2 to 10^5 Ωm volume resistivity: almost as good an insulating material as, say, flexible PVC. Raw water that might be taken from the mains water supply and water used in the circulation system of central heating systems that contains additives may well be laden with sufficient impurities to make the water conductive to some extent.

The potential referred to in the definition is generally taken to be that of the conductive mass of Earth. This is by convention taken as 0 V.

Metal pipework and other extraneous-conductive-parts, such as steel stanchions, that enter the equipotential zone are generally considered to be at 0 V.

However, it is not just earthy extraneous-conductive-parts that the designer needs to consider. A potential related to the earth fault in another building may be imported along an extraneous-conductive-part such as pipework.

Even when an extraneous-conductive-part is raised to a potential, it may not be capable of introducing its potential to a person (or livestock) if it cannot be touched by a person in simultaneous contact with another conductive part, such as an exposed-conductive-part or another extraneous-conductive-part at a different potential, or a live part.

Consequently, for a conductive part to be able to introduce a potential to a person (or livestock) in contact with the potential of any of the conductive parts, that part should be accessible to such a person (or livestock).

On occasions, a conductive part may be partially insulated from Earth potential by virtue of some insulating material intervening between the conductive part and earth. This might be the case, for example, where an insulating board used as part of the building construction places a high resistance between the two points. In such cases, Equation (6.1) should be used:

$$R_{CP} > \left(\frac{U_0}{I_B}\right) - Z_T \qquad (6.1)$$

where:

R_{CP} is the measured resistance between the conductive part concerned and the MET of the installation (in ohms).

U_0 is the nominal voltage to Earth of the installation (in volts).

I_B is the value of current through the human body (or livestock) which should not be exceeded (in amperes).

Z_T is the total impedance of the human body or livestock (in ohms).

Where the measured resistance R_{CP} satisfies the value calculated using Equation (6.1), the conductive part should not be considered to be an extraneous-conductive-part. However, the designer's decision should also take into account the likely stability of the resistance of the conductive part over the lifetime of the installation.

BSI publication BS IEC 60479-1:2018 *Effects of current on human beings and livestock. General aspects* provides data for Z_T and I_B. The value of Z_T differs from person to person and is also dependent on a number of factors, including (but not limited to) the touch voltage, the supply frequency, the duration of the current flow, the conditions of wetness of the skin and the surface area of contact. For the onerous current path of one hand to feet and water-wet conditions where U_0 is 230 V at 50 Hz, it is not unreasonable to take a value of 1000 Ω for Z_T for the purposes of this exercise. The designer can then select a suitable value of I_B, such as:

▶ 0.5 mA — the threshold of perception;
▶ 10 mA — the threshold of let-go; or
▶ 30 mA — a current that can cause effects such as the following, depending on the time within which automatic disconnection occurs.
▶ where automatic disconnection occurs within 300 ms (for example, by non-delay RCD with $I_{\Delta n}$ not exceeding 30 mA): perception and involuntary muscular contractions, but usually no harmful electrical physiological effects;

> ▶ longer disconnection time, up to 5 s: involuntary muscular contractions, difficulty in breathing, reversible disturbances of heart function and immobilization, but usually no organic damage.

Using Equation (6.1) to illustrate the range of values of resistance R_{CP} for currents from 0.5 mA, 10 mA and 30 mA, we get Equations (6.2),(6.3) and (6.4), respectively:

$$R_{CP} > \left(\frac{U_0}{I_B}\right) - Z_T = \left(\frac{230}{0.5 \times 10^{-3}}\right) - 1000 = 460\,000 - 1000 = 459 \text{ k}\Omega \qquad (6.2)$$

$$R_{CP} > \left(\frac{U_0}{I_B}\right) - Z_T = \left(\frac{230}{10 \times 10^{-3}}\right) - 1000 = 23\,000 - 1000 = 22 \text{ k}\Omega \qquad (6.3)$$

$$R_{CP} > \left(\frac{U_0}{I_B}\right) - Z_T = \left(\frac{230}{30 \times 10^{-3}}\right) - 1000 = 7667 - 1000 = 6.67 \text{ k}\Omega \qquad (6.4)$$

It can thus be seen from Equation (6.4) that if the designer elects to accept 30 mA as a safe level and the resistance of the extraneous-conductive-part to the MET (R_{CP}) is above the threshold of 6.67 kΩ, then the conductive part need not be considered to be an extraneous-conductive-part, provided that automatic disconnection of the supply to the relevant circuit occurs within 300 ms.

The definition of extraneous-conductive-part does not include conductive parts that form part of the electrical installation.

6.2 Some examples of extraneous-conductive-parts

411.3.1.2 Regulation 411.3.1.2 of BS 7671 provides a list of parts that may meet the definition of an extraneous-conductive-part:

▶ water installation pipes;
▶ gas installation pipes;
▶ other installation pipework and ducting;
▶ central heating and air conditioning systems; and
▶ exposed metallic structural parts of a building.

With the introduction of BS 7671:2018, any metallic pipe entering the building having an insulating section at its point of entry need not be connected to the protective equipotential bonding.

The connection of a lightning protection system to the protective equipotential bonding is required to be made in accordance with BS EN 62305.

This list should not be regarded as exhaustive. Figure 6.1 shows a typical layout of extraneous-conductive-parts and associated main protective bonding conductors.

▼ **Figure 6.1** Typical protective equipotential bonding

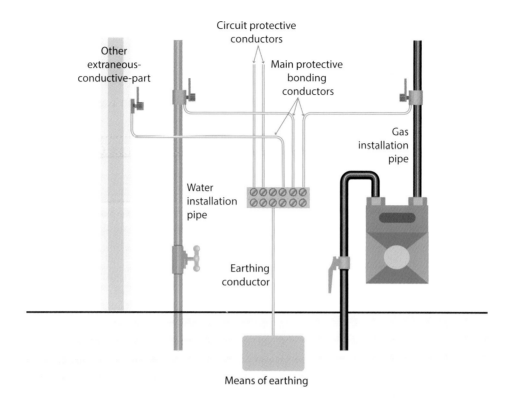

Means of earthing

6.3   An example of a conductive part which is not an extraneous-conductive-part

701.415.2   Some insight into the subject of conductive parts that are not extraneous-conductive-parts can be gained from Regulation 701.415.2, relating to supplementary bonding in a location containing a bath or shower. Indent (iii) of the regulation states that metallic door architraves, window frames and similar parts are not considered to be extraneous-conductive-parts unless they are connected to metallic structural parts of the building.

6.4   Connection to pipework

BS 951:2009 *Electrical earthing. Clamps for earthing and bonding. Specification* is the performance specification for clamps for connection of main protective and supplementary bonding conductors to pipes, solid rods and the like, with circular cross-sections. The specification embraces the connection of:

▶ earthing conductors having a csa in the range 2.5 mm^2 to 70 mm^2 to earth electrode rods or other means of earthing; and

▶ bonding conductors to metal tubes of circular cross-section that have circumferences of not less than 18.8 mm (i.e. diameters of not less than 6 mm).

It is important to note that BS 951 clamps are not intended, and are therefore not suitable, for connection to the armouring or lead sheath of a cable. Clamps applied to an armoured cable can have the effect of crushing the bedding or insulation, thus reducing its effectiveness. Similarly, a clamp fitted to a lead cable sheath can cause the lead to cold-flow and, with continuing expansion and contraction expected under varying load conditions, this often results in a high-resistance connection. This, in turn, will increase the earth fault loop impedance and may adversely affect disconnection times.

Nor is the clamp intended for any use other than to encircle a pipe or rod; for example, it should not be used to connect to an item that is too large for the clamp.

BS 951 lays down important mechanical constraints for metal clamps that are used for the purpose of connecting conductors to provide mechanically and electrically sound earthing and bonding connections to metal tubes.

Figure 6.2 shows a correctly installed BS 951 clamp connecting a copper protective conductor and a copper tube. The slots in the label are intended as an aid to packaging and storage only.

▼ **Figure 6.2**   BS 951 clamp

BS 951 calls for the clamp to consist of:

(a) a device for making electrical contact with the tube;
(b) a means of tightening the device on to the tube;
(c) a means of locking the arrangement given in item (b); and
(d) a termination separate from the arrangement given in item (b) for attaching the protective conductor to the device given in item (a).

From the above, it should be noted that the clamp for the protective conductor is separate from the means of tightening and locking the clamp to make electrical contact with the tube.

A screw termination is capable of accepting one of the following:

▶   a conductor clamped under a screw head provided with a captive washer so that the screw head does not act directly on the conductor (also capable of accepting a looped unbroken conductor); or
▶   a single conductor clamped directly by a screw-threaded arrangement, the csa of the conductor being within the range specified in Table 1 of BS 951 (also capable of accepting a looped unbroken conductor); or
▶   a bolted-on cable socket from a range of sockets that can accommodate conductors having a csa covering the whole range specified in Table 1 of BS 951.

Table 6.1 replicates the data relating to termination reference and conductor csa given in Table 1 of BS 951:2009.

▼ **Table 6.1** Termination reference and conductor size

Termination reference	Nominal csa of conductor (mm²)
A	2.5
B	4
C	6
D	10
E	16
F	25
G	35
H	50
I	70

522.5.1 Clamps are made of different metals in order to suit different environments and the consequential level of corrosion. Regulation 522.5.1 of BS 7671 demands that such clamps are suitably protected from the effects of corrosion, or manufactured from material that is resistant to corrosion.

514.13.1 Figure 6.3 illustrates three clamps made from different metals and differing lengths, together with the warning notice required at each clamp by Regulation 514.13.1, with the words 'SAFETY ELECTRICAL CONNECTION – DO NOT REMOVE'.

Some manufacturers make BS 951 clamps available with colour coding to denote the environmental conditions for which it is claimed they are suitable. The manufacturer's advice regarding the colour coding scheme should be followed.

▼ **Figure 6.3** Three clamps of different metals and differing lengths

522.5.2
522.5.3
Regulations 522.5.2 and 522.5.3 require that metals liable to initiate electrolytic action should not be placed in contact with one another. This would be relevant to clamps and their contact with a dissimilar metal and, in particular, to the aluminium warning notice label, which is liable to severe corrosion where in contact with other metals.

526.3
With certain exceptions, all connections and joints are required to be accessible for inspection, testing and maintenance purposes (see Regulation 526.3). Earthing and bonding clamps are no exception to this requirement, and the clamp and its conductor termination are required to be accessible. Additionally, such clamps are required to be complete with their warning label, as shown in Figures 6.2 and 6.4.

▼ **Figure 6.4** BS 951 earthing and bonding clamp warning label

BS 951 clamps may not be able to accommodate pipes and other circular components with a large csa, which are sometimes encountered in the construction and services of buildings. The answer is not to join two or more BS 951 clamps together, but to use a suitable proprietary clamp, as shown in Figure 6.5.

▼ **Figure 6.5** Clamps for larger diameter pipes etc.

6.5 Connections to structural steelwork and buried steel grids

BS 951 clamps are suitable for use with pipes and rods with a circular csa. However, extraneous-conductive-parts often exist in the form of irregular shapes that are required to be used as protective bonding conductors and/or used in supplementary equipotential bonding. Suitable clamps are commercially available from suppliers; however, the installation designer must only use clamps manufactured from materials that will not adversely affect the extraneous-conductive-part to which they connect. Table 11 of BS 7430:2011+A1:2015 provides recommendations for the manufacture of earthing components.

526.1
526.2
As with all connections, these clamps must meet the requirements of Regulation 526.1, which states that every connection must provide durable electrical continuity and adequate mechanical strength. Regulation 526.2 meanwhile, requires that the connection be suitable for the conductors. The clamp should therefore be suitable for connection to the extraneous-conductive-part, in terms of its shape, csa and dimensions, as well as being able to act as a supplementary bonding conductor. The clamp should be equipped with a means of locking to prevent loosening due to, forexample, vibration.

Figure 6.6 illustrates some typical examples of clamps for connecting main protective bonding conductors to steel building components.

▼ **Figure 6.6** Typical examples of clamps for connecting to steel building components

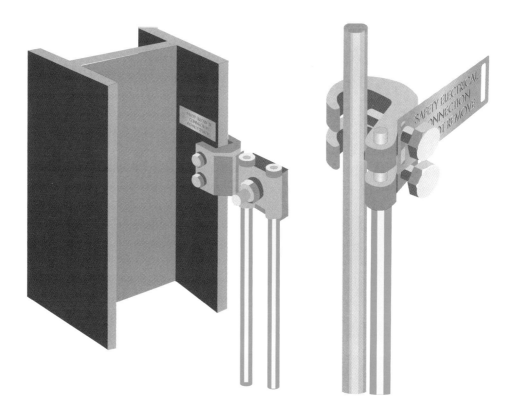

Automatic disconnection

7.1 Automatic disconnection of supply (ADS)

411.1(ii) Automatic disconnection of supply is a protective measure that provides fault protection by virtue of coordinating the characteristics of the protective device for automatic disconnection and the relevant impedance of the circuit concerned. To achieve this, BS 7671 requires:

- installation equipment metalwork (exposed-conductive-parts) to be connected to the earthing arrangement;
- protective equipotential bonding; and
- automatic disconnection of the supply under earth fault conditions.

Whatever the type of system, the operating characteristic of each protective device for automatic disconnection and the earth fault loop impedance of the associated circuit should be properly coordinated so as to achieve automatic disconnection within the prescribed time limit set out in BS 7671.

411.3.2.1 In order to meet the prescribed time limits for automatic disconnection, Regulation 411.3.2.1 requires coordination between:

- the time/current characteristic of each protective device for automatic disconnection;
- the earthing arrangements; and
- the relevant impedance of the circuit concerned.

This co-ordination is required to make certain that under earth fault conditions the potentials between simultaneously accessible exposed-conductive-parts and extraneous-conductive-parts are of a magnitude and duration such as not to cause danger.

7.2 TN systems

411.4.4 For TN systems, automatic disconnection within a specified time is fulfilled when Equation (7.1), given in Regulation 411.4.4, is satisfied:

$$Z_s \times I_a \leq U_0 \times C_{min} \quad (V) \tag{7.1}$$

where:

Z_s is the impedance in ohms (Ω) of the fault loop comprising the source, the line conductor up to the point of the fault, and the protective conductor between the point of the fault and the source

I_a is the current in amperes (A) causing the automatic operation of the disconnecting device within the time specified in Table 41.1 or, as appropriate, Regulation 411.3.2.3. Where an RCD is used, this current is the rated residual operating current providing disconnection in the time specified in Table 41.1 or Regulation 411.3.2.3.

U_0 is the nominal AC rms or ripple-free DC line voltage to Earth in volts (V).

C_{min} is the minimum voltage factor to take account of voltage variations depending on time and place, changing of transformer taps and other considerations.

Note: For a low voltage supply given in accordance with the ESQCR, C_{min} is given the value 0.95.

Table 41.1
411.3.2.2 For convenience, data contained in Table 41.1 of BS 7671 for AC is replicated here in Table 7.1.

▼ **Table 7.1 Table 41.1 of BS 7671: Maximum disconnection times for TN systems (see Regulation 411.3.2.2)**

Installation nominal voltage, U_0 (V AC) line voltage to Earth	Maximum disconnection time, t (s)
120	0.8
230	0.4
400	0.2
> 400	0.1

Notes:
(a) For voltages that are within the supply tolerance band (230 + 10 % −6 %), the disconnection time appropriate to the nominal voltage applies.
(b) For intermediate values of voltage, the higher value of the voltage range in the table is to be used.

Table 7.1 sets out the prescribed time limits for automatic disconnection in TN systems for final circuits with a rated current not exceeding:

(a) 63 A with one or more socket-outlets; and
(b) 32 A supplying only fixed connected current-using equipment.

The prescribed time limits for automatic disconnection in a TN system set out in Table 7.1 do **not** apply to:

▶ a distribution circuit;
▶ a final circuit not covered in (a) or (b), immediately above;
411.8 ▶ reduced low voltage circuits (as described in Regulation 411.8); and
Sect 714 ▶ street lighting circuits (as described in Section 714).

411.3.2.3
411.3.2.4 For TN and TT systems, for those circuits and equipment for which Table 41.1 does not apply, a disconnection time of not more than 5 s and 1 s respectively is permitted.

7.2.1 Earth fault loop impedance

Table 41.2
Table 41.3
Table 41.4 For a circuit with a nominal voltage to Earth (U_0) of 230 V, the earth fault loop impedance values given in Tables 41.2 to 41.4 of BS 7671 appropriate to the required disconnection time may be used for the types and ratings of the over current devices listed.

Appx 3 For convenience, Tables 7.2 to 7.4 give the limiting earth fault loop impedances for different disconnection times and for a number of common overcurrent protective devices. The circuit loop impedances in the tables have been determined using a value for C_{min} of 0.95. The tables also include 80 % 'rule of thumb' values for comparison with measured values of the earth fault loop impedance, Z_s, as described in Appendix 3 of BS 7671. This is necessary where the measurements are taken at an ambient temperature of around 20 °C and where the normal operating temperature for the conductors is 70 °C. Figure 7.1 shows the relationship between conductor resistance and temperature. Equation (7.2) allows an assessment to be made of the likely increase in conductor resistance from that at an ambient temperature of 20 °C to normal operating temperature of 70 °C.

▼ **Figure 7.1** Conductor resistance against temperature

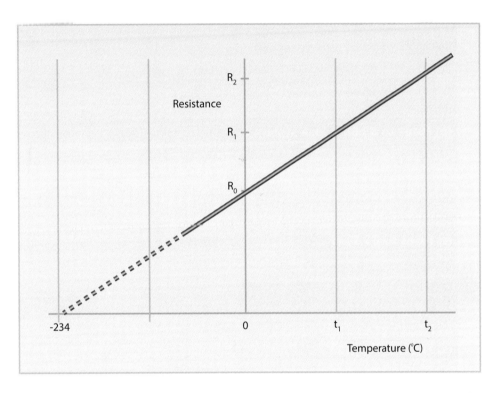

Equation (7.2) allows the resistance to be assessed at an elevated temperature t if it is known at a lower temperature:

$$R_t = R_{20}[1 + \alpha_{20}(t - 20)] (\Omega) \qquad (7.2)$$

where:

R_{20}   is the conductor resistance at 20 °C.

R_t is the conductor resistance at temperature t.

α_{20} is the resistance/temperature coefficient at 20 °C.

To establish the resistance of a copper conductor of 2 Ω resistance at a temperature of 20 °C when at a temperature of 70 °C, we apply Equation (7.2); the resistance/temperature coefficient at 20 °C for copper is 0.004 per °C:

$$R_t = R_{20}[1+\alpha_{20}(t-20)] = 2[1+0.004(70-20)] = 2[1+0.2] = 2.4\,(\Omega) \qquad (7.3)$$

Equation (7.3) shows that the resistance at 70 °C is 20 % higher than it is at 20 °C. Bearing in mind that reactance is unaffected by temperature, this would support the 80 % values given for comparison with measured values of earth fault loop impedance, Z_s, as a rule of thumb.

Table 41.2
411.3.2.2

▼ **Table 7.2**   Data from Table 41.2 of BS 7671 for maximum earth fault loop impedance (Z_s) for fuses, for 0.4 s disconnection time with U0 of 230 V (see Regulation 411.3.2.2)[1], together with 80 % values[2] for comparison with measured values taken at an ambient temperature of 20 °C

(a) General purpose (gG) and motor circuit (gM) fuses to BS 88-2 – fuse systems E (bolted) and G (clip in)

Rating (A)	2	4	6	10	16	20	25	32
Z_s (Ω)	33.10	15.60	7.80	4.65	2.43	1.68	1.29	0.99
Z_s (Ω) measured[2]	26.48	12.48	6.24	3.72	1.94	1.34	1.03	0.79

(b) Fuses to BS 88-3 fuse system C

Rating (A)	5	16	20	32	45	63
Z_s (Ω)	9.93	2.30	1.93	0.91	0.57	0.36
Z_s (Ω) measured[2]	7.94	1.84	1.54	0.73	0.46	0.29

(c) Fuses to BS 3036

Rating (A)	5	15	20	30
Z_s (Ω)	9.10	2.43	1.68	1.04
Z_s (Ω) measured[2]	7.28	1.94	1.34	0.83

(d) Fuses to BS 1362

Rating (A)	3	13
Z_s (Ω)	15.60	2.30
Z_s (Ω) measured[2]	12.48	1.84

Table 41.3 ▼ **Table 7.3**   Data from Table 41.3 of BS 7671 for maximum earth fault loop impedance
411.3.2.2   (Z_s) for circuit-breakers with U0 of 230 V, for operation giving compliance
411.3.2.3   with the 0.4 s disconnection time of Regulation 411.3.2.2 and the 5 s
                             disconnection time of Regulation 411.3.2.3[1], together with 80 % values[2] for
                             comparison with measured values taken at an ambient temperature of 20 °C

(a) Type B circuit-breakers to BS EN 60898 and the overcurrent characteristics of RCBOs to BS EN 61009													
Rating (A)	3	6	10	16	20	25	32	40	50	63	80	100	125	I_n
Z_s (Ω)	14.57	7.28	4.37	2.73	2.19	1.75	1.37	1.09	0.87	0.69	0.55	0.44	0.35	230 × 0.95/ (5 I_n)
Z_s (Ω) measured[2]	11.66	5.82	3.50	2.18	1.75	1.40	1.10	0.87	0.70	0.55	0.44	0.35	0.28	

(b) Type C circuit-breakers to BS EN 60898 and the overcurrent characteristics of RCBOs to BS EN 61009													
Rating (A)		6	10	16	20	25	32	40	50	63	80	100	125	I_n
Z_s (Ω)		3.64	2.19	1.37	1.09	0.87	0.68	0.55	0.44	0.35	0.27	0.22	0.17	230 × 0.95/(10I_n)
Z_s (Ω) measured[2]		2.91	1.75	1.10	0.87	0.70	0.54	0.44	0.35	0.28	0.22	0.18	0.14	

(c) Type D circuit-breakers to BS EN 60898 and the overcurrent characteristics of RCBOs to BS EN 61009													
Rating (A)		6	10	16	20	25	32	40	50	63	80	100	125	I_n
Z_s (Ω), 0.4 s		1.82	1.09	0.68	0.55	0.44	0.34	0.27	0.22	0.17	0.14	0.11	0.09	230 × 0.95/ (20I_n)
Z_s (Ω), 0.4 s measured[2]		1.46	0.87	0.54	0.44	0.35	0.27	0.22	0.18	0.14	0.11	0.09	0.07	
Z_s (Ω), 5 s		3.64	2.19	1.37	1.09	0.87	0.68	0.55	0.44	0.35	0.27	0.22	0.17	230 × 0.95/(10I_n)
Z_s (Ω), 5 s measured[2]		2.91	1.75	1.10	0.87	0.70	0.54	0.44	0.35	0.28	0.22	0.18	0.14	

Table 41.4 ▼**Table 7.4**   Data from Table 41.4 of BS 7671 for maximum earth fault loop impedance (Z_s)
411.3.2.3   for 5 s disconnection time with U0 of 230 V (see Regulation 411.3.2.3)[1],
                             together with 80 % values[2] for comparison with measured values taken at an
                             ambient temperature of 20 °C

(a) General purpose (gG) and motor circuit application (gM) fuses to BS 88-2 – fuse systems E (bolted) and G (clip in)									
Rating (A)	2	4	6	10	16	20	25	32	40
Z_s (Ω)	44.0	21.0	12.0	6.80	4.00	2.80	2.20	1.7	1.30
Z_s (Ω) measured[2]	35.2	16.8	9.60	5.44	3.20	2.24	1.76	1.36	1.04
Rating (A)	50	63	80	100	125	160	200		
Z_s (Ω)	0.99	0.78	0.55	0.42	0.32	0.27	0.18		
Z_s (Ω) measured[2]	0.79	0.62	0.44	0.34	0.26	0.22	0.14		
(b) Fuses to BS 88-3 fuse system C									
Rating (A)	5	16	20	32	45	63	80	100	

Z_s (Ω)	14.6	3.90	3.20	1.60	1.00	0.68	0.51	0.38
Z_s (Ω) measured[2]	11.7	3.12	2.56	1.28	0.80	0.54	0.41	0.30

(c) Fuses to BS 3036

Rating (A)	5	15	20	30	45	60	100
Z_s (Ω)	16.8	5.08	3.64	2.51	1.51	1.07	0.51
Z_s (Ω) measured[2]	13.4	4.06	2.91	2.01	1.21	0.86	0.41

(d) Fuses to BS 1362

Rating (A)	3	13
Z_s (Ω)	22.0	3.64
Z_s (Ω) measured[2]	17.60	2.91

For a circuit with a nominal voltage other than 230 V, it is important to recognise that for the operation of an overcurrent protective device, such as a fuse or circuit-breaker, it is the same magnitude of the earth fault current that is required. Therefore, taking a practical example of a circuit with a nominal voltage (U_0) of 115 V protected by a 16 A, BS 88-2 fuse supplying socket-outlets, the limit on the earth fault loop impedance (Z_s) would be the value given in Table 7.2 or Table 41.2 of BS 7671 (2.43 Ω) multiplied by a factor of 115/230 (1/2), representing the ratio of two different values of U_0. The limit on Z_s would therefore be 1.22 Ω with the circuit line conductor at its maximum permitted operating temperature, and the circuit protective conductor at the appropriate assumed initial temperature, as given in Tables 54.2 to 54.5 of BS 7671.

The earth fault loop can be seen to have a number of discrete constituent parts which all contribute to the earth fault loop impedance, as shown in Figure 7.2, from which the principal parts shown are:

▶ source impedance;
▶ supply line conductor;
▶ installation line conductor;
▶ circuit protective conductor;
▶ earthing conductor; and
▶ PEN or PE conductor of the supply.

For the general case, the earth fault loop impedance (Z_s) is as given in Equation (7.4):

$$Z_s = Z_e + (Z_1 + Z_2)(\,\Omega\,) \tag{7.4}$$

where:

Z_e is the external earth fault loop impedance (made up principally of the source impedance, the supply line conductor and the PEN conductor of the supply).

Z_1 is the impedance of the installation line conductor to the point of the fault.

Z_2 is the impedance of the circuit protective conductor (cpc).

By convention, the impedance of the fault is taken to be negligible.

For circuits rated at less than 100 A and operating at a supply frequency not exceeding

50 Hz, the inductive reactance can be ignored. Therefore, for most final circuits, the impedances Z_1 and Z_2 may be replaced by the resistive components of impedance R_1 and R_2, the resistances of the line and the CPC, as given in Equation (7.5):

$$Z_s = Z_e + (R_1 + R_2)(\Omega)$$
(7.5)

where:

Z_e is the external earth fault loop impedance (made up principally of the source impedance, the supply line conductor and the PEN of the supply).

R_1 is the resistance of the installation line conductor to the point of the fault.

R_2 is the resistance of the CPC.

Table 41.2
Table 41.3
Table 41.4
To make certain that automatic disconnection occurs within the specified time, the value of Z_s at the furthest (the electrically most remote) point of a circuit of nominal voltage to earth (U_0) of 230 V should not exceed the limiting value given in Tables 41.2 to 41.4 of BS 7671 (Tables 7.2 to 7.4 of this Guidance Note) for the selected protective device and for the prescribed limit on disconnection time. The limiting values given in the tables assume that the conductors are at their normal operating temperature. Where the conductors are at a different temperature when the values of the impedances or resistances are established (normally by measurement), the readings should be adjusted accordingly.

▼ **Figure 7.2** Earth fault loop (TN-C-S system)

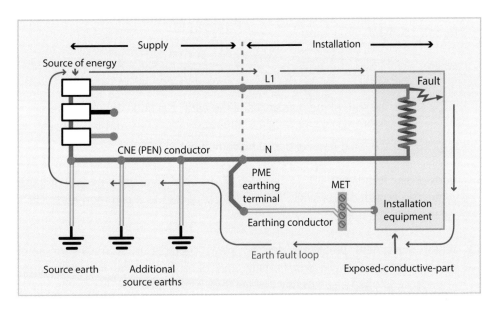

7.2.2 Automatic disconnection using an RCD in TN systems

411.4.5 An RCD may be used to provide fault protection in TN systems where the earth fault loop impedance (Z_s) is insufficiently low to operate an overcurrent protective device (fuse or circuit-breaker) within the prescribed disconnection time.

411.4.4 Where an RCD is used to provide automatic disconnection, the application of Regulation 411.4.4 requires Equation (7.6) to be met:

$$Z_s \times I_a \times \leq U_0 \times C_{min} (V)$$
(7.6)

where:

Z_s is the impedance in ohms (Ω) of the fault loop comprising the source, the line conductor up to the point of the fault, and the protective conductor between the point of the fault and the source.

I_a is the current in amperes (A) causing the automatic operation of the disconnecting device within the time specified in Table 41.1 of Regulation 411.3.2.2 or, as appropriate, Regulation 411.3.2.3. Where an RCD is used, this current is the rated residual operating current providing disconnection in the time specified in Table 41.1 or Regulation 411.3.2.3.

C_{min} is the minimum voltage factor to take account of voltage variations depending on time and place, changing of transformer taps and other considerations.

Note: For a low voltage supply given in accordance with the ESQCR, C_{min} is given the value 0.95.

Where more than one residual current device is used in series, the upstream device should be time-delayed or S-type and the downstream a type for general use, both to BS EN 61008 or BS EN 61009 to achieve discrimination. This is addressed in clause 7.6.

7.3 TT systems

411.5.3 For installations forming part of a TT system, the limits on automatic disconnection are once again given in terms of time (Table 7.5) and the following Equation (7.7) (refer to Regulation 411.5.3):

$$R_A \times I_{\Delta n} \leq 50\,V \tag{7.7}$$

where:

R_A is the sum of the resistances of the earth electrode and the protective conductor(s) connecting it to the exposed-conductive-parts (Ω).

$I_{\Delta n}$ is the rated residual operating current of the RCD (A).

411.5.1 Additionally, Regulation 411.5.1 requires all exposed-conductive-parts that have fault protection provided by a single RCD to be connected to a common earth electrode, via the MET.

411.5.2 Whilst overcurrent protective devices are acceptable as a means of automatic disconnection in installations forming part of a TT system, the earth fault loop impedance is often insufficiently low for their widespread use; hence the preference in Regulation 411.5.2 for an RCD to be used.

Table 7.6, which replicates information given in Table 41.5 of BS 7671, provides maximum earth fault loop impedances (Z_s) to ensure RCD operation for final circuits not exceeding:

(i) 63 A with one or more socket-outlets; and
(ii) 32 A supplying only fixed connected current-using equipment.

Where protection against earth fault conditions is provided by an RCD and the Z_s values of Tables 41.2 to 41.4 are not achieved but do meet Table 41.5 values, the requirements of Section 434 must be met.

▼ **Table 7.5** Table 41.1 of BS 7671: Maximum disconnection times – extract for TT systems (see Regulation 411.3.2.2)

Installation nominal voltage, U_0 (V AC)	Maximum disconnection time, t (s)
120	0.3
230	0.2
400	0.07
> 400	0.04

Notes:
(a) For voltages that are within the supply tolerance band (230 +10 % −6 %), the disconnection time appropriate to the nominal voltage applies.
(b) For intermediate values of voltage, the higher value of the voltage range in the table is to be used.

▼ **Table 7.6** Table 41.5 of BS 7671: Maximum earth fault loop impedance (Z_s) for non-delayed and time-delayed 'S' Type RCDs to BS EN 61008-1 and BS EN 61009-1 for U0 of 230 V (See Regulation 411.5.3.)

Rated residual operating current (mA)	Maximum earth fault loop impedance, Z_s (ohms)
30	1667*
100	500*
300	167
500	100

Notes:
▶ Figures for Z_s result from the application of Regulation 411.5.3(i) and (ii). Disconnection must be ensured within the times stated in Table 7.5. Table 7.5 is based on Table 41.1.
* The resistance of the installation earth electrode should be as low as practicable. A value exceeding 200 Ω may not be stable. Refer to Regulation 542.2.4.

7.4 IT systems

The source of an IT system is either isolated from Earth or, where necessary to reduce overvoltage or to damp voltage oscillations, earthed through a high impedance.

411.6.2 Disconnection in the event of a first line fault to Earth is not essential, but precautions must be taken to safeguard against the risk of electric shock in the event of two faults existing simultaneously. All exposed-conductive-parts are required to be earthed and the condition expressed in Regulation 411.6.2 and in Equation (7.8) for AC should be met:

$$R_A \times I_d \leq 50 \, V \tag{7.8}$$

where:

> R_A is the sum of the resistances of the earth electrode and the protective conductor connecting it to the exposed-conductive-parts.
>
> I_d is the fault current of the first fault of negligible impedance between a line conductor and an exposed-conductive-part.

For DC, the voltage limit in Equation (7.8) becomes 120 V.

411.6.4 The fault current I_d takes account of leakage currents and the total earthing impedance of the electrical installation. As called for in Regulation 411.6.4, it is necessary to provide either an insulation monitoring device (IMD) or a residual current monitor (RCM), capable of providing a visual and/or audible warning of the occurrence of a first fault from a live part to an exposed-conductive-part or to Earth.

On the occurrence of a second fault, the system should be treated as if it were a TN system or a TT system, depending on how the exposed-conductive-parts are earthed:

411.6.5
- ▶ where the exposed-conductive-parts are earthed collectively, the conditions for a TN system should apply, but subject to Regulation 411.6.5; and
- ▶ where exposed-conductive-parts are earthed in groups or individually, conditions for a TT system should apply.

For cases in which the neutral is not distributed (three-phase three-wire distribution), Equation (7.9) applies. Where the neutral is distributed (three-phase four-wire distribution and single-phase distribution), Equation (7.10) applies:

$$Z_s \leq \frac{U \times C_{min}}{2I_a} \left(\Omega \right) \tag{7.9}$$

$$Z_s^1 \leq \frac{U_0 \times C_{min}}{2I_a} \left(\Omega \right) \tag{7.10}$$

where:

> U is the nominal AC rms or DC voltage, in volts, between line conductors.
>
> U_0 is the nominal AC rms or DC voltage, in volts, between a line conductor and the neutral conductor or midpoint conductor, as appropriate.
>
> Z_s is the impedance in ohms of the fault loop comprising the line conductor and the protective conductor of the circuit.
>
> Z_s^1 is the impedance in ohms of the fault loop comprising the neutral conductor and the protective conductor of the circuit.

Table 41.1 I_a is the current in amperes (A) causing operation of the protective device within the time specified in Regulation 411.3.2.2 for TN systems or, as appropriate, Regulation 411.3.2.3..

> C_{min} is the minimum voltage factor to take account of voltage variations depending on time and place, changing of transformer taps and other considerations.

Note: For a low voltage supply given in accordance with the ESQCR, C_{min} is given the value 0.95.

▼ **Table 7.7**   Table 41.1 of BS 7671: Maximum disconnection time in IT systems (second fault)

Installation nominal voltage U_0 (V)	Maximum disconnection time t (s)
$50 < U_0 \leq 120$	0.8
$120 < U_0 < 230$	0.4
$230 < U_0 \leq 400$	0.2
$U_0 > 400$	0.1

Note: U_0 is the nominal AC rms voltage between line and neutral conductors.

Both overcurrent protective devices and residual current protective devices are permitted to be used for automatic disconnection. However, where fault protection is provided by an RCD, each final circuit should be separately protected.

7.5   Additional protection

411.3.3   For all socket-outlets in AC systems with a rated current not exceeding 32 A, and mobile equipment with a rated current not exceeding 32 A for use outdoors, additional protection by means of an RCD shall be provided as required by Regulation 411.3.3. An exception is permitted for socket-outlets, other than for an installation in a dwelling, where a documented risk assessment determines that RCD protection is not necessary.

Note that these requirements do not apply to functional extra-low voltage (FELV) or reduced low voltage systems.

415.1.1   An RCD selected for additional protection is required to have a rated residual operating current, $I_{\Delta n}$, of not more than 30 mA. The RCD provides additional protection in the event of failure of basic protection or fault protection or carelessness by users.

415.1.2   An RCD is not permitted to be used as the sole means of protection against electric shock (see Regulation 415.1.2).

7.6   RCDs in series

536.4.1.4   It is often necessary to employ RCDs in series, or cascade them as, for example, in an
314.1   installation that forms part of a TT system. Regulation 536.4.1.4 of BS 7671 states requirements for coordinating the characteristics of the RCDs such that the intended selectivity is achieved. Even where it is not strictly necessary to provide selectivity to prevent danger or to minimize inconvenience (Regulation 314.1), it is considered good practice to do so.

For two RCDs in series, selectivity will generally be achieved by selecting a time-delayed characteristic for the upstream RCD. Figure 7.3 shows two RCDs in series feeding an item of Class I current-using equipment that has developed an earth fault,

▼ **Figure 7.3** RCDs in series feeding an item of Class I current-using equipment

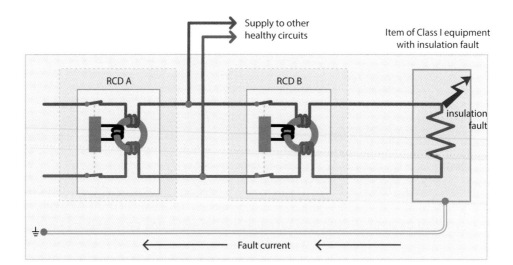

As a consequence of the earth fault, an earth fault current, I_{fault}, flows through the line conductor and on to earth via the protective conductor. This earth fault current, by flowing in the protective conductor, causes an imbalance in the currents in the live conductors. This imbalance is sensed in both RCD A and RCD B. So, where neither device has any time-delay mechanism, both devices are likely to operate.

This operation of both devices is likely even where there is a significant difference in the rated residual currents of the devices. Where this occurs, the supply could be lost to healthy circuits as well as causing disconnection of the circuit with the earth fault.

It is important to recognise that an RCD does not limit the earth fault current, which is a function of the nominal voltage to Earth, U_0, and the earth fault loop impedance, Z_s. The minimum earth fault current is given in Equation (7.11):

$$I_{fault} = \frac{U_0 \times C_{min}}{Z_s} (A)$$ (7.11)

Unless the earth fault loop impedance, Z_s, is unusually (and probably unacceptably) high, the earth fault current will exceed that required to operate both devices. Take, for example, a situation in which the earth fault loop impedance is 100 Ω (TT system). Assuming a value of 0.95 for C_{min}, the resulting earth fault current, I_{fault}, would be 2.18 A, as shown by Equation (7.12):

$$I_{fault} = \frac{U_0 \times C_{min}}{Z_s} = \frac{230 \times 0.95}{100} = 2.18\,A$$ (7.12)

An earth fault current of 2.18 A is likely to exceed the rated residual operating currents of both devices and therefore both will operate. To discriminate in terms of magnitude only is not therefore a plausible option in most cases.

Selectivity between RCDs can generally only be achieved by 'forcing' a delay in the operation of the upstream RCD, thus allowing the downstream RCD time to disconnect the fault before the time delay of the upstream RCD has been exhausted.

Guidance Note 8: Earthing and Bonding
© The Institution of Engineering and Technology

Where an RCD is provided for additional protection, a time-delay mechanism is not permitted. Such devices are required to meet the type-test disconnection time laid down in the product standard.

Where an RCD is provided for fault protection, and does have a time-delay mechanism, it is required to automatically disconnect within the specified overall time limit for the particular circuit(s).

7.7 Automatic disconnection for reduced low voltage system

A reduced low voltage system is often used where, for functional reasons, the use of protective extra-low voltage (PELV) is impracticable and separated extra-low voltage (SELV) is not necessary.

Part 2 BS 7671 defines a *reduced low voltage system* as:

> *A system in which the nominal line-to-line voltage does not exceed 110 volts and the nominal line to Earth voltage does not exceed 63.5 volts.*

411.8.4.1 Although the source may be via a motor-generator or a source independent of other supplies, such as an engine-driven generator, the most common source is a double-wound isolating transformer complying with BS EN 61558-1 (formerly BS 3535) and BS EN 61558-2-23.

Figure 7.4 shows the secondary windings of both a three-phase and a single-phase source.

▼ **Figure 7.4** Sources for reduced low voltage systems

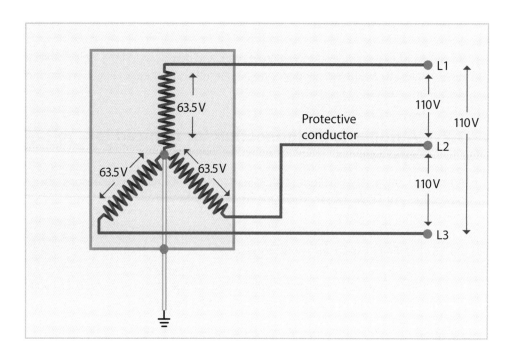

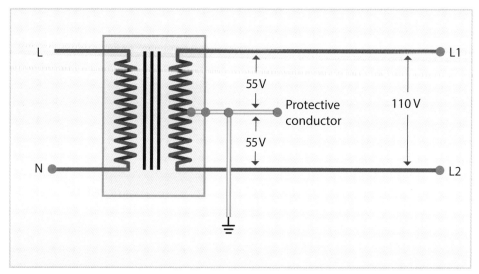

411.8.3 As required by Regulation 411.8.3, fault protection is to be provided by automatic disconnection. Automatic disconnection is not reliant on earthed equipotential bonding and for this reason it is commonly-used as a protective measure for hand-held tools and small mobile plant on construction and demolition sites. However, it is important that all exposed-conductive-parts are connected through protective conductors to Earth.

A reduced low voltage system, which requires the star-point of a three-phase system or the midpoint of a single-phase system to be earthed, is not to be confused with 110 V systems often used with control circuitry, which are not so connected. The latter can only be described as low voltage and the general requirements of BS 7671 apply.

Care is required in the selection of a portable generator for a reduced low voltage system. Many 240/110 V single-phase generators do not have a centre-tap available and are therefore likely to be earthed on one pole only. This would not constitute a reduced low voltage system as prescribed in BS 7671, and would not provide the same degree of fault protection. It would, instead, be a low voltage system.

411.8.3 Regulation 411.8.3 states that where automatic disconnection is provided by overcurrent protective devices (such as a fuse or a circuit-breaker), a device is required in each line conductor. Disconnection for fault protection is required to occur in not more than 5 s, and this applies to every point in the circuit, including socket-outlets. For such devices, the earth fault loop impedance, Z_s, is required to meet that given in Equation (7.13):

$$Z_s \leq \frac{U_0 \times C_{min}}{I_a}\left(\Omega\right)$$

(7.13)

where:

Z_s is the impedance in ohms (Ω) of the fault loop comprising the source, the line conductor up to the point of the fault, and the protective conductor between the point of the fault and the source.

I_a is the current in amperes (A) causing the automatic operation of the disconnecting device within the time specified in Table 41.1 or, as appropriate, Regulation 411.3.2.3. Where an RCD is used, this current is the rated residual operating current providing disconnection in the time specified in Table 41.1 or Regulation 411.3.2.3.

U_0 is the nominal AC rms or DC line voltage to Earth in volts (V).

C_{min} is the minimum voltage factor to take account of voltage variations depending on time and place, changing of transformer taps and other considerations.

Note: For a low voltage supply given in accordance with the ESQCR, C_{min} is given the value 0.95.

The limiting value of the earth fault loop impedance, Z_s, may be calculated using Equation (7.13). Alternatively, Z_s can be determined from Table 41.6 of BS 7671, the data of which is replicated here as Table 7.8.

Table 41.6 ▼ **Table 7.8**   Table 41.6 of BS 7671: Maximum earth fault loop impedance (Z_s) for a
411.8.1.2   disconnection time of 5 s and U_0 of 55 V (single-phase) or 63.5 V (three-
411.8.3   phase) (see Regulations 411.8.1.2 and 411.8.3)

	Circuit-breakers to BS EN 60898 and the overcurrent characteristics of RCBOs to BS EN 61009-1				General purpose (gG) fuses to BS 88-2 – fuse systems E and G	
	Type B		Types C and D			
Un (Volts)	55	63.5	55	63.5	55	63.5
Rating amperes			Z_s (ohms)			
3	3.48	4.02	1.74	2.01		
6	1.74	2.01	0.87	1.01	2.90	3.35
10	1.05	1.21	0.52	0.60	1.63	1.89
16	0.65	0.75	0.33	0.38	0.95	1.10
20	0.52	0.60	0.26	0.30	0.67	0.77
25	0.42	0.48	0.21	0.24	0.52	0.60

	Circuit-breakers to BS EN 60898 and the overcurrent characteristics of RCBOs to BS EN 61009-1				General purpose (gG) fuses to BS 88-2 – fuse systems E and G	
	Type B		**Types C and D**			
U_0 (Volts)	55	63.5	55	63.5	55	63.5
Rating amperes	Z_s (ohms)					
32	0.33	0.38	0.16	0.19	0.42	0.48
40	0.26	0.30	0.13	0.15	0.31	0.35
50	0.21	0.24	0.10	0.12	0.24	0.27
63	0.17	0.19	0.08	0.10	0.19	0.22
80	0.13	0.15	0.07	0.08	0.13	0.15
100	0.10	0.12	0.05	0.06	0.12	0.14
125	0.08	0.10	0.04	0.05	0.08	0.09
I_n	$\dfrac{10.4}{I_n}$	$\dfrac{12.1}{I_n}$	$\dfrac{5.2}{I_n}$	$\dfrac{6.1}{I_n}$		

Notes:

1 The circuit loop impedances have been determined using a value for factor C_{min} of 0.95.

2 The circuit loop impedances given in the table should not be exceeded when:
   (i) the line conductors are at the appropriate maximum permitted operating temperature, as given in Table 52.1 of BS 7671; and
   (ii) the circuit protective conductors are at the appropriate assumed initial temperature, as given in Tables 54.2 to 54.5 of BS 7671. If the conductors are at a different temperature when tested, the reading should be adjusted accordingly. See IET Guidance Note 1 or Appendix 3 of BS 7671.

3 Where the line conductor insulation is of a type for which Table 52.1 gives a maximum permitted operating temperature exceeding 70 °C, such as thermosetting, but the conductor has been sized in accordance with Regulation 512.1.5:
   (i) the maximum permitted operating temperature for the purpose of note 2(i) is 70 °C; and
   (ii) the assumed initial temperature for the purpose of note 2(ii) is that given in Tables 54.2 to 54.4 of BS 7671 corresponding to an insulation material of 70 °C thermoplastic.

4 Data for fuses of rating exceeding 200 A should be obtained from the manufacturer.

411.8.3 Where an RCD is used to provide automatic disconnection, Regulation 411.8.3 requires that the product of the rated residual operating current in amperes, $I_{\Delta n}$, and the earth fault loop impedance, Z_s, in ohms does not exceed 50 V, as expressed in Equation (7.14):

$$I_{\Delta n} \times Z_s \leq 50\,V \tag{7.14}$$

The earth fault loop impedance, Z_s, may be assessed by the use of Equation (7.15).

$$Z_s = Z_p \times \left(\frac{V_s}{V_p}\right)^2 + \left(\frac{Z\% \text{ Trans} \times V_s^2}{100 \times VA}\right) + (R_1 + R_2)_s \qquad (7.15)$$

where:

Z_p is the loop impedance of the primary circuit including that of the source of supply, Z_e.

$Z\% \text{ Trans}$ is the percentage impedance of the step-down transformer.

VA is the rating of the step-down transformer.

V_s is the secondary voltage.

V_p is the primary voltage.

$(R_1 + R_2)_s$ are the secondary circuit line and protective conductor resistances.

Where data on the step-down transformer is not available, Equation (7.15) may be simplified to the following:

$$Z_s = \left(Z_p \times \left(\frac{V_s}{V_p}\right)^2 + (R_1 + R_2)_s\right) \qquad (7.16)$$

i.e.

$$Z_s = \left(\left[\left(Z_e + (2R_1)_p\right) \times \left(\frac{V_s}{V_p}\right)^2\right] + (R_1 + R_2)_s\right) \qquad (7.17)$$

where:

Z_e is the external line-neutral/earth loop impedance.

$(2R_1)_p$ is the primary circuit line plus neutral conductor impedance.

7.8 Automatic disconnection and alternative supplies

313.1
551.4.1
An important aspect of providing fault protection, which can be readily overlooked by the designer, is the need to ensure satisfactory operation of the relevant protective device(s) when the installation, or part thereof, is energized from a second source. This may be an alternative low voltage feed, a standby generator supply or an uninterruptible power supply (UPS). To be certain that the requirements of BS 7671 for protection against electric shock (and short-circuits) will still be satisfied, the designer must obtain full information for the alternative or special supply and make the necessary checks on his/her design, which may have been based only upon the characteristics of the normal supply source.

A generator control panel or piece of UPS equipment may include self-protection, a feature of which is the rapid collapse of the output voltage to the load. This will inhibit the operation of any fault protective device situated beyond the equipment terminals and the feature cannot be assumed to provide a fail-safe operational arrangement for the user. Safety of the system as a whole should be ensured by, if necessary, involving the equipment supplier.

Sect 414 ## 7.9 Separated extra-low voltage systems (SELV)

Part 2 By definition, *SELV* is:

> *An extra-low voltage system which is electrically separated from Earth and from other systems in such a way that a single fault cannot give rise to the risk of electric shock.*

It should be appreciated that if a single fault were to occur, it is intended that this should be confined to the SELV system and not involve any conductive parts of another system. Thus, the construction of a SELV system necessitates the use of high-integrity equipment and materials. Figure 7.5 shows the configuration of a SELV source, from which it is noted that the secondary winding has no connection with Earth.

▼ **Figure 7.5** A SELV source

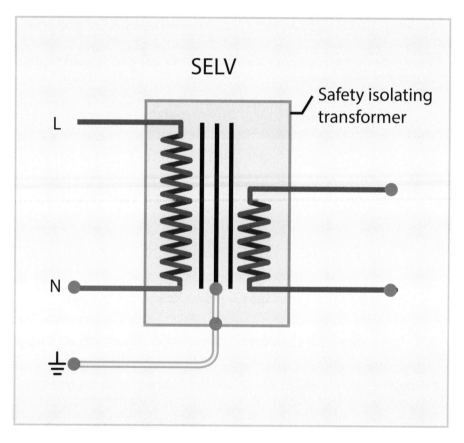

A SELV system is an extra-low voltage (ELV) system that is electrically separated from Earth and from other systems and is itself a protective measure that is deemed to provide both basic protection and fault protection.

Sect 414 ## 7.10 Protective extra-low voltage systems (PELV)

The degree of safety afforded by a SELV system depends crucially upon it being isolated both from Earth and from any other system. If this cannot be achieved and maintained throughout its life, the extra-low voltage system cannot depend solely upon the SELV.

Part 2 By definition, *PELV* is:

> *An extra-low voltage system which is not electrically separated from Earth, but which otherwise satisfies all the requirements for SELV.*

Figure 7.6 shows the configuration of a PELV source, from which it is noted that the secondary winding has a connection with Earth.

▼ **Figure 7.6** A PELV source

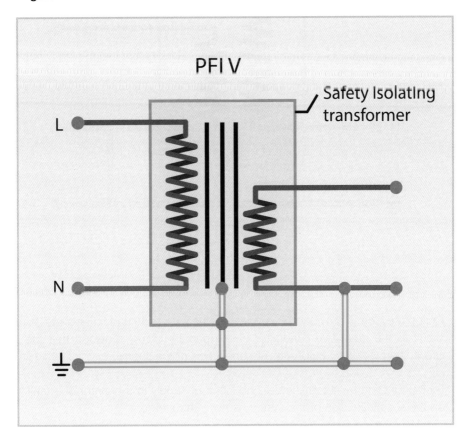

PELV systems rely upon the primary circuit to clear faults introduced from that circuit. It should be confirmed that this fault protection is appropriate for the location.

Faults elsewhere in the installation will introduce fault voltages into the PELV system via the protective conductor.

411.7
Part 2
7.11 Functional extra-low voltage systems (FELV)

Functional extra-low voltage (FELV) is used where extra-low voltage is required for functional purposes, such as machine control systems. Figure 7.7 shows the configuration of a FELV source, from which it is noted that the secondary winding has a connection with Earth. The source may have only simple separation by basic insulation or may be similar to PELV.

▼ **Figure 7.7** A FELV source

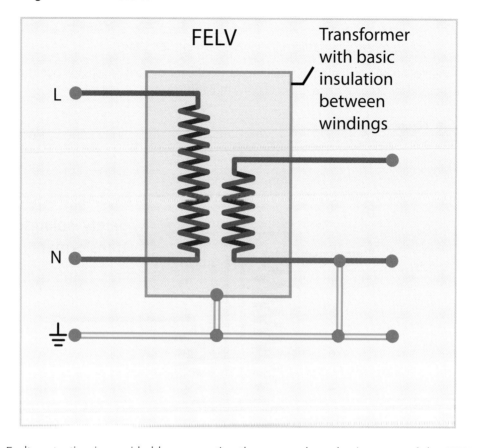

411.7.3 Fault protection is provided by connecting the exposed-conductive-parts of the FELV circuit to the protective conductor of the primary circuit, provided that the primary circuit is protected by automatic disconnection of the supply.

Fault protection by automatic disconnection of supply in the event of a fault to earth on the secondary side of the system is not a requirement.

7.12 Overvoltages imposed on the low voltage (LV) system due to high voltage (HV) system earth faults

442.2 It is a requirement of Section 422 of BS 7671 that the low voltage installation designer considers stress voltages on the earthing arrangement of the low voltage supply caused by earth faults in the high voltage system.

Where the substation is installed, owned and operated by the distributor in accordance with the requirements of the ESQCR and the Distribution Code, then Section 442 is of no consequence to the designer of the low voltage installation.

Should a privately owned substation be installed, then the requirements of Section 442 should be consulted in conjunction with those of BS EN 61936-1:2010+A1:2014 *Power installations exceeding 1 kV a.c. Common rules*, and BS EN 50522:2010 *Incorporating corrigendum October 2012, Earthing of power installations exceeding 1 kV a.c.*

BS EN 50522, Table 2, gives minimum requirements for determining whether it is feasible to interconnect high voltage and low voltage earthing arrangements. Feasibility is dependent on the earthing design for the substation achieving minimum safe touch and step voltages and a tolerable 'earth potential rise' (EPR – stress voltage).

The IET Electrical Installation Design Guide gives guidance on the calculations involved in applying the requirements of Section 442.

Guidance Note 8: Earthing and Bonding
© The Institution of Engineering and Technology

Supplementary protective equipotential bonding

8

8.1 The purpose of supplementary protective equipotential bonding

Supplementary protective equipotential bonding is considered as an addition to the fault protection required by BS 7671 in certain situations and locations. Supplementary protective equipotential bonding involves connecting together the exposed-conductive-parts and extraneous-conductive-parts, such as those shown in Figure 8.1, to minimize the touch voltages between them under earth fault conditions of a circuit. The figure shows both protective equipotential bonding and supplementary protective equipotential bonding. It is important to note that protective equipotential bonding is always required where the protective measure is automatic disconnection of supply (ADS).

Where necessary, supplementary protective equipotential bonding is applied to an installation or location in order to establish an equipotential zone within that particular part of the installation or location. It has the effect of re-establishing the equipotential reference at that location for all the exposed-conductive-parts and extraneous-conductive-parts that are required to be bonded together locally. This further reduces any potential differences that may arise between any of these parts during an earth fault.

419.3 Supplementary protective equipotential bonding is required by BS 7671 to be provided in the following circumstances:

- ▶ in installations and at locations of increased shock risk, some of which are addressed in Part 7 of BS 7671; and
419.3 ▶ where, in the event of an earth fault, the conditions for automatic disconnection cannot be fulfilled in the time required by Regulation 411.3.2.2, 411.3.2.3 or 411.3.2.4, as appropriate.

415.2 Where supplementary protective equipotential bonding is required, compliance with
544.2 Regulations 415.2 and 544.2.1 to 544.2.5, as appropriate, is necessary.

Figure 8.1 shows two items of current-using equipment ('A' and 'B') and the cpcs of the circuits feeding them, together with two separate extraneous-conductive-parts ('C' and 'D'). In accordance with the requirements for protective equipotential bonding in BS 7671, the cpcs and the extraneous-conductive-parts are connected to the MET of the installation.

544.2.1 Supplementary protective equipotential bonding is carried out by installing bonding
544.2.2 conductors between the exposed-conductive-parts and the extraneous-conductive-parts, by making the connections shown between items of current-using equipment A and B and between B and F. The conductors connecting A with B are required to comply with Regulation 544.2.1, whereas the conductor between B and extraneous-conductive-part C (connection made at point F) is required to comply with Regulation 544.2.2.

▼ **Figure 8.1** Application of supplementary protective equipotential bonding in a special location

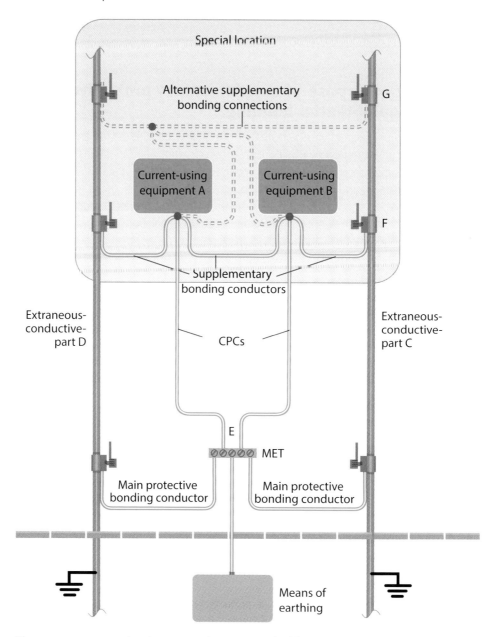

544.2.3 The extraneous-conductive-part D is connected either to current-using equipment A
544.2.4 or B or to extraneous-conductive-part C, as indicated by the broken lines. Where it is connected to either item of current-using equipment, the bonding conductor is required to comply with Regulation 544.2.2. Alternatively, where the connection is made instead to the extraneous-conductive-part C, compliance with Regulation 544.2.3 is required. It should be noted that, as permitted by Regulation 544.2.4, the portion of extraneous-conductive-part C between the points and G can be considered part of the supplementary protective equipotential bonding.

It is not a requirement of BS 7671 to connect the supplementary bonding conductor back to the MET of the installation, although as shown in Figure 8.1, the locally bonded parts will be connected to this terminal by virtue of one or more cpcs and/or extraneous-conductive-parts.

419.3 As previously mentioned, supplementary protective equipotential bonding may be required in locations where there is considered to be an increased risk of electric shock. Additionally, it may also be required where the conditions for automatic disconnection cannot be fulfilled (see Regulation 419.3). The locations and situations where supplementary protective equipotential bonding is required by BS 7671 are addressed in clauses 8.6 to 8.14.

8.2   Supplementary bonding conductor types

543.2.2
543.2.5 Although a single-core, non-flexible, copper cable having a green-and-yellow covering is often used as a supplementary equipotential bonding conductor, there are other types that are also suitable for such use, as permitted by Regulations 543.2.2 and 543.2.5, including:

- ▶ a protective conductor forming part of a sheathed cable;
- ▶ metal parts of wiring systems; and
543.2.3 ▶ extraneous-conductive-parts (for example, metallic hot and cold water pipes and heating pipework). Note that neither a gas nor an oil pipe may be so used. (See Regulation 543.2.3.)

544.2.5 Where supplementary protective equipotential bonding is required to a Class I fixed appliance connected by a short length of flexible cable from an adjacent connection unit or flex outlet accessory, Regulation 544.2.5 permits this to be provided by the cpc within the flexible cable. However, the cpc is required to be connected to the supplementary protective equipotential bonding within the connection unit or other accessory. Figure 8.2 indicates the application of this regulation.

▼ **Figure 8.2** Supplementary protective equipotential bonding via a short flexible cable

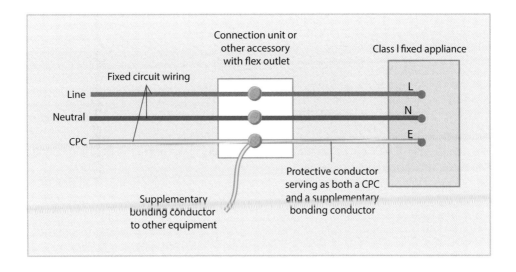

8.3 Csa of supplementary bonding conductors

544.2 The requirements for the minimum csa of supplementary bonding conductors are given in Regulations 544.2.1 to 544.2.5, which are summarized in Table 8.1.

544.2.1
544.2.2 Where the supplementary bonding conductor referred to in Table 8.1 is made from a different metal to that of the cpc, the minimum csa is determined by the application of Equation (8.1) (where Regulation 544.2.1 applies – connecting together two exposed-conductive-parts) or Equation (8.2) (where Regulation 544.2.2 applies – connecting together one exposed-conductive-part and one extraneous-conductive-part):

$$S_b \geq S_p \left(\frac{\rho b}{\rho p} \right) (mm^2) \tag{8.1}$$

$$S_b \geq \frac{S_p}{2} \left(\frac{\rho b}{\rho p} \right) (mm^2) \tag{8.2}$$

where:

S_b is the minimum csa (mm²) required for the bonding conductor.

S_p is the csa (mm²) of the cpc.

ρb is the resistivity (Ωm) for the conductor material of the supplementary bonding conductor.

ρp is the resistivity (Ωm) for the conductor material of the cpc.

▼ **Table 8.1** Minimum cross-sectional area of supplementary bonding conductors

Connecting conductive parts	Conductor type	Csa of conductors
Two exposed-conductive-parts (Regulation 544.2.1)	Where sheathed[1] or otherwise provided with mechanical protection.	For conductors of the same metal, the bonding conductor to have a conductance not less than that of the smaller cpc connected to the exposed-conductive-parts. For other metals, see below.
	Where not provided with mechanical protection (e.g. by a sheath or enclosure in conduit etc.).	Not less than 4 mm².
Exposed-conductive-part to an extraneous-conductive-part (Regulation 544.2.2)	Where sheathed[1] or otherwise provided with mechanical protection.	For conductors of the same metal, the bonding conductor to have a conductance not less than half that of the protective conductor connected to the exposed-conductive-part. For other metals, see below.
	Where not provided with mechanical protection (e.g. by a sheath or enclosure in conduit etc.).	Not less than 4 mm².

Connecting conductive parts	Conductor type	Csa of conductors
Two extraneous-conductive-parts (Regulation 544.2.3)	Where neither of the extraneous-conductive-parts is connected to an exposed-conductive-part[2].	Not less than: 2.5 mm² if sheathed[1] or otherwise provided with mechanical protection, or 4 mm² if not provided with mechanical protection.

Notes:

1 The insulation of a single-core non-sheathed cable (e.g. to BS 6004 or BS 7211) is not a sheath.

2 Where one of the extraneous-conductive-parts is connected to an exposed-conductive-part, Regulation 544.2.2 is to be applied to the supplementary bonding conductor connecting the two extraneous-conductive-parts (see details earlier in this table).

8.4 Limitations on resistance of supplementary bonding conductors

415.2.2 In AC systems, Regulation 415.2.2 requires that the resistance, R, of the supplementary bonding conductor connecting exposed-conductive-parts and extraneous-conductive-parts is limited to the value expressed in Equation (8.3).

$$R \leq \frac{50\,V}{I_a}\left(\Omega\right) \tag{8.3}$$

where: for the case of an overcurrent protective device, I_a is the minimum current that disconnects the circuit within 5 s, whereas for the case of an RCD, I_a is the rated residual operating current $I_{\Delta n}$.

To determine the limiting resistance of supplementary bonding conductors for a particular location, all the circuits in that location have to be considered using the value of the highest I_a or $I_{\Delta n}$ in the formula. For example, consider a circuit protected by a 45 A, BS 88-3 fuse system C with a minimum current (I_a) required to disconnect in 5 s of 220 A. Inputting this data into Equation (8.3) gives a limit on the resistance of the supplementary bonding conductor of 0.22 Ω:

$$R \leq \frac{50V}{I_a} = \frac{50}{220} = 0.22\,\Omega \tag{8.4}$$

Using the conductor resistance for 4 mm² copper of 4.61 mΩ/m and with a limit on the resistance of 0.22 Ω, the supplementary bonding conductor could be as long as 48 m. For a 2.5 mm² copper conductor at 7.41 mΩ/m, the limit on resistance of 0.22 Ω would limit the conductor length to 30 m.

For most if not all circumstances, this limitation on the resistance, R, will not result in a need to increase the csa of the supplementary bonding conductors above 4 mm² copper. However, this constraint will need to be checked for the most onerous circuit for each special location.

8.5 Supports for supplementary bonding conductors

Where a supplementary bonding conductor forms part of a composite cable, such as a separate core, the method of supporting the bonding conductor will be dictated by the type of cable and the manufacturer's installation instructions.

543.3.1 Single-core supplementary bonding conductors, and similar conductors, are required to be adequately supported, avoiding non-electrical services such as pipework as a means of support, so that they can resist external influences such as mechanical damage(see Section 522 of BS 7671) and any anticipated factors likely to result in deterioration (see Regulation 543.3.1).

Table 8.2 provides some guidance in providing adequate support for supplementary bonding conductors. Figure 8.3 shows an example of a supplementary bonding conductor properly supported.

▼ **Table 8.2** Recommended spacing of supports of single-core rigid copper cables for supplementary bonding conductors

Overall cable diameter, d (mm)	Horizontal spacing (mm)	Vertical spacing (mm)	Comment
d ≤ 3	100	150	This data is provided as a guide and may be overridden by the constraints of good workmanship and by visual considerations
3 < d ≤ 5	150	200	
5 < d < 10	200	250	
10 < d < 15	300	350	

▼ **Figure 8.3** An example of supports for a supplementary bonding conductor

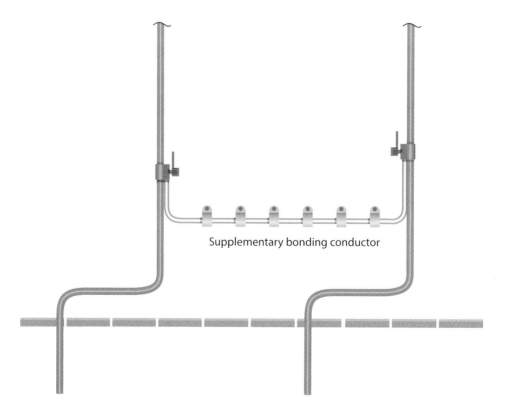

Supplementary bonding conductor

8.6 Locations containing a bath or shower

Locations containing a bath and/or shower present an increased risk of electric shock due to:

▶ a reduction in body resistance occasioned either by bodily immersion or by wet skin; and
▶ likely contact of substantial areas of the body with earth potential.

Sect 701 Section 701 in Part 7 of BS 7671, 'Special installations or locations', specifically addresses the particular requirements for such locations. However, it is important to note that these requirements modify or supplement the general requirements contained elsewhere in BS 7671. In other words, compliance with all the relevant requirements of BS 7671 is essential, not just Section 701.

415.2 The general requirements for additional protection by supplementary protective
701.415.2 equipotential bonding are given in Regulations 415.2.1 and 415.2.2. For this location, these regulations are supplemented by Regulation 701.415.2.

701.415.2 Except where certain conditions are met, as explained later in this section, Regulation 701.415.2 requires that local supplementary protective equipotential bonding is provided, connecting together the terminals of the protective conductor of each circuit supplying Class I and Class II equipment to the accessible extraneous-conductive-parts within a location containing a bath or shower. In so doing, the occurrence of voltages between any such parts of a magnitude sufficient to cause danger of electric shock is prevented. BS 7671 identifies a number of parts which may be considered to be extraneous-conductive-parts:

▶ metallic pipes supplying services, and metallic waste pipes (for example, water, gas);
▶ metallic central heating pipes and air conditioning systems; and
▶ accessible metallic structural parts of the building (metallic door architraves, window frames and similar parts are not considered to be extraneous-conductive-parts unless they are connected to metallic structural parts of the building).

This list of extraneous-conductive-parts is not exhaustive; any other parts that fall within the definition of an extraneous-conductive-part given in Part 2 of BS 7671 will require supplementary protective equipotential bonding.

701.415.2 It is also a requirement of Regulation 701.415.2 that the necessary supplementary bonding is carried out either within the location containing the bath or shower or in close proximity. In this context, close proximity is taken to mean in an adjoining roof void or an airing cupboard opening into, or adjoining, the room containing the bath or shower. The regulation expresses a preference for the point of connection of the supplementary bonding to be close to the point of entry of the service (extraneous-conductive-part) into the location.

The requirements for supplementary protective equipotential bonding in this location apply to accessible extraneous-conductive-parts.

414.4.4 Regulation 414.4.4 requires that supplementary protective equipotential bonding is *not* carried out to exposed-conductive-parts of SELV circuits.

Figures 8.4 and 8.5 show sectional views of a typical bathroom, the first with metal pipework and the second with non-metallic pipework, where the option of supplementary bonding has been selected.

▼ **Figure 8.4** Supplementary protective equipotential bonding in a bathroom – metal pipe installation

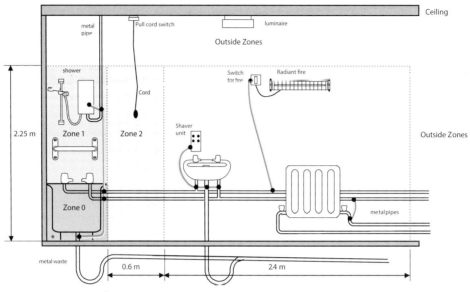

* Zone 1 if the space is accessible without the use of a tool.
Spaces under the bath, accessible only with the use of a tool, are outside the zones.

Notes:

1 The protective conductors of all power and lighting points within the zones must be supplementary bonded to all extraneous-conductive-parts in the zones, including metal waste, water and central heating pipes.

2 Circuit protective conductors may be used as supplementary bonding conductors.

As mentioned earlier, supplementary bonding of the terminals of cpcs of circuits supplying Class II equipment is required, as shown in Figures 8.4 and 8.5. This permits Class II equipment to be replaced by Class I equipment during the lifetime of the installation, without the need for further supplementary bonding which might be difficult and costly to carry out subsequently.

▼ Figure 8.5 Supplementary protective equipotential bonding in a bathroom – non-metallic pipe installation

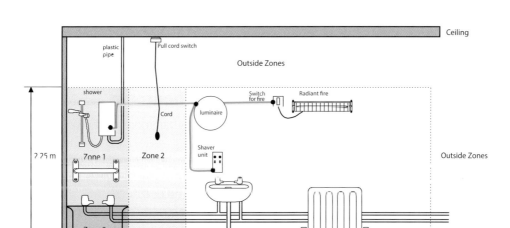

* Zone 1 if the space is accessible without the use of a tool.
Spaces under the bath, accessible only with the use of a tool, are outside the zones.

415.2.2 For this special location, Regulation 415.2.2 requires that the resistance, R, between simultaneously accessible exposed-conductive-parts and extraneous-conductive-parts shall fulfil the condition expressed in Equation (8.5):

$$R \leq \frac{50V}{I_a} \left(\Omega \right) \qquad (8.5)$$

where: for the case of an overcurrent protective device, I_a is the operating current that disconnects the circuit within 5 s, whereas for the case of an RCD, I_a is the rated residual operating current $I_{\Delta n}$.

Guidance on the csa and limiting resistance of supplementary bonding conductors is given in clauses 8.3 and 8.4, respectively.

701.415.2 Regulation 701.415.2 of BS 7671 permits supplementary protective equipotential bonding in a location containing a bath or shower to be omitted in a building where the following three conditions are met:

(a) all final circuits of the location comply with the requirements for automatic disconnection;

(b) all final circuits of the location have additional protection by means of an RCD with a rated residual operating current $(I_{\Delta n})$ not exceeding 30 mA; and

(c) all extraneous-conductive-parts of the location are effectively connected to the protective equipotential bonding in accordance with Regulation 411.3.1.2.

701.512.3
415.1 Low voltage socket-outlets (230 V) within a location containing a bath or shower have been permitted since BS 7671:2008 came into effect, but they have to be protected by an RCD having a rated residual operating current, $I_{\Delta n}$, not exceeding 30 mA, in accordance with Regulation 415.1, and they must not be installed within 3 m horizontally from the boundary of zone 1.

701.753 Where electric heating is embedded in the floor of a bathroom or shower room, Regulation 701.753 requires that the heating elements are covered by a fine mesh metallic grid or have a metallic sheath, connected in either case to the local supplementary protective equipotential bonding. Even if the heating elements with a metallic sheath are covered by an earthed grid, the heating elements are required to be earthed in the normal way.

8.7 Shower cabinet located in a bedroom

The requirements of BS 7671 are the same as for a bathroom or shower room.

701.512.3
415.1.1 Low voltage socket-outlets (for example, 230 V) may be installed not closer than 3 m horizontally from the boundary of zone 1, provided they are protected by an RCD having a rated residual operating current, $I_{\Delta n}$, not exceeding 30 mA, in accordance with Regulation 415.1.1. In cases of retrofit, this requirement could be met by the provision of a socket-outlet with an integral residual current device, as shown in Figure 8.6. Alternatively, the circuit could be protected with a residual-current circuit-breaker with overcurrent protection (i.e. an RCBO).

▼ **Figure 8.6** Socket-outlet with RCD ($I_{\Delta n}$ = 30 mA)

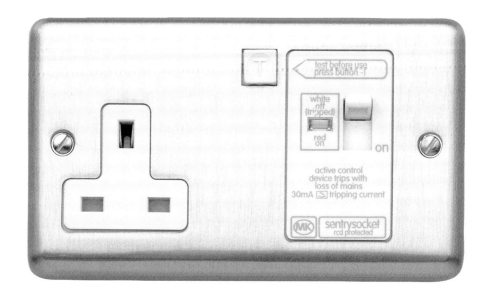

8.8 Swimming pools and other basins

Swimming pools and the basins of fountains and paddling pools are locations where there is an increased risk of electric shock due to:

▶ a reduction in body resistance occasioned either by bodily immersion or by wet skin; and
▶ likely contact of substantial areas of the body with earth potential.

Sect 702 Section 702 in Part 7 of BS 7671, 'Swimming pools and other basins', specifically addresses the particular requirements for such locations. However, it is important to note that these requirements modify or supplement the general requirements contained elsewhere in BS 7671. In other words, compliance with all the relevant requirements of BS 7671, not just Section 702, is essential.

As with locations containing a bath or shower, the zonal concept is used for swimming pools, as shown in Figures 8.7 and 8.8 (which replicate Figures 702.1 and 702.2 of BS 7671). In the case of swimming pools, the zones are designated 0, 1 and 2 (the dimensions for which are given in Figures 8.7 and 8.8).

Fig 702.1 ▼ **Figure 8.7** Zone dimensions for swimming pools and paddling pools

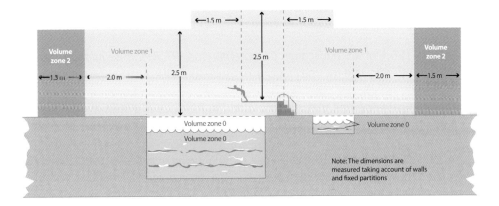

Fig 702.2 ▼ **Figure 8.8** Zone dimensions for a basin above ground

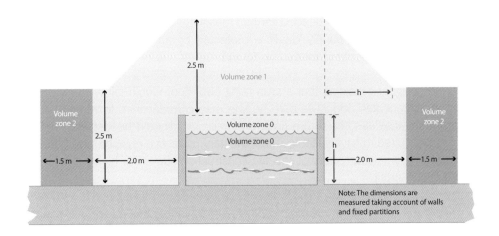

702.415.2 Regulation 702.415.2 requires supplementary protective equipotential bonding to
414.4.4 connect all extraneous-conductive-parts in zones 0, 1 and 2 with the protective conductors of all exposed-conductive-parts of equipment situated in these zones. However, bonding is *not* to be carried out to exposed-conductive-parts of SELV circuits.

Typically, parts which may fall within the definition of an extraneous-conductive-part would include:

▶ exposed structural steelwork;
▶ metal pipes etc.; and
▶ metal handrails.

Guidance on the csa and limiting resistance of supplementary bonding conductors is given in clauses 8.3 and 8.4, respectively.

702.410.3.4.3 Where a metallic grid is installed in a solid floor in zone 2, and such a grid meets the definition of an extraneous-conductive-part, supplementary protective equipotential bonding should be provided to connect it with other extraneous-conductive-parts and the protective conductors of all exposed-conductive-parts in the zones.

702.410.3.4.3(ii) note It is recommended that where a swimming pool (or basin) installation is connected to a PME earth, an earth mat or earth electrode of suitably low resistance is connected to the PME earth terminal. A metallic grid may perform as such an electrode. BS 7671 advises that a resistance to true earth of 20 ohms or less is appropriate.

Section 11.5.4 of IET publication *Commentary on IET Wiring Regulations* provides an equation to calculate the maximum resistance to earth.

702.55.1 An electric heating unit may be embedded in the floor of a swimming pool, provided that either:

▶ it is protected by SELV, the source of SELV being installed outside zones 0, 1 and 2. However, it is permitted to install the source of SELV in zone 2 if its supply circuit is additionally protected by an RCD having the characteristics specified in Regulation 415.1.1; or

▶ it incorporates an earthed metallic sheath connected to the supplementary bonding and its supply circuit is protected by an RCD having a rated residual operating current not exceeding 30 mA; or

▶ it is covered by an embedded earthed metallic grid connected to the supplementary bonding and its supply circuit is protected by an RCD having a rated residual operating current not exceeding 30 mA.

If the designer considers that a swimming pool floor would fall within the definition of an extraneous-conductive-part, he/she may consider that, to remedy a problem that may arise at some future time, a metallic grid should be installed. The metallic grid is required to have suitable supplementary protective equipotential bonding. This, of course, is relatively easy at the site construction stage, but highly problematic and costly to retrofit.

Figure 8.9 illustrates an example of providing supplementary protective equipotential bonding to a metallic grid. Conductors and connections, which should be accessible for inspection, testing and maintenance, should be selected to withstand all the adverse effects of the external influences likely to be present, such as corrosion and mechanical damage.

▼ **Figure 8.9** A means of supplementary protective equipotential bonding a metallic grid

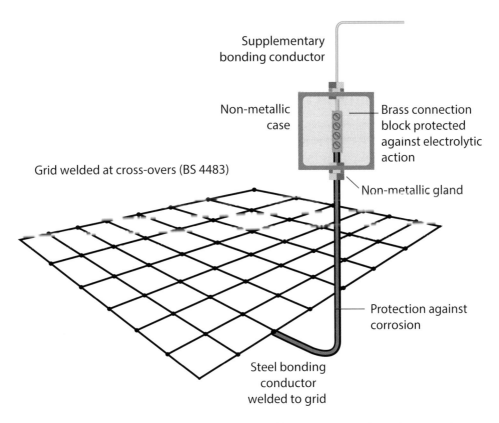

526.3   Where connections cannot be made in accessible positions, such as those made to a metal grid within a concrete slab, they are required by Regulation 526.3 to be made by:

- ▶ welding;
- ▶ soldering; or
- ▶ brazing; or
- ▶ an appropriate compression tool.

At least two connections should be made to the metallic grid, preferably at two or more diagonally extreme points. Additionally, for the metallic grid to be entirely effective, all constituent parts of the grid should be reliably and durably connected together.

In addition to electric shock, as defined, there is also the perception of electric shock to consider in such locations. *Electric shock* is defined as:

Part 2   *A dangerous physiological effect resulting from the passing of an electric current through a human body or livestock.*

541.2   The magnitude of the current through the human body that is discernible varies from person to person, but it is generally accepted that perception of an electric current can occur from 0.5 mA. With this in mind, the installation designer may decide that the swimming pool installation should form part of a TT system rather than make use of the electricity distributor's earthing facility. In other words, the installation would be earthed to an installation earth electrode rather than to the supply earthing terminal. This may overcome a possible problem with discernible voltages associated in the supply neutral transmitted to the extraneous-conductive-parts of the installation.

However, this solution is not without difficulties. The exposed-conductive-parts and the extraneous-conductive-parts earthed to the installation earth electrode of the special location must be electrically separated from parts of the site that are earthed to the distributor's earthing facility. Simultaneous inaccessibility should be maintained at all times.

8.9 Agricultural and horticultural premises

Sect 705   The particular requirements for agricultural and horticultural premises are given in Section 705 of BS 7671 and apply to all parts of fixed installations indoors and outdoors. They also apply to locations where livestock is kept and to storage areas and the like, such as:

- ▶ stables;
- ▶ chicken-houses;
- ▶ piggeries;
- ▶ feed-processing locations;
- ▶ lofts; and
- ▶ storage areas for hay, straw and fertilizers.

Locations within the curtilage of agricultural and horticultural premises, intended solely for human habitation, are excluded from the scope of this section, although they might have within them requirements for supplementary bonding.

It is important to note that the requirements of Section 705 modify or supplement the general requirements contained elsewhere in BS 7671. In other words, compliance with all the relevant requirements of BS 7671 is essential, not just Section 705

Agricultural and horticultural premises are locations where there is perceived to be an increased risk of electric shock to humans and livestock due to:

- ▶ a reduction in body resistance occasioned either by partial bodily immersion or by wet skin or hides; and
- ▶ likely contact of substantial areas of the body with earth potential.

415.2   The general requirements for supplementary protective equipotential bonding are
544.2   given in Regulations 415.2 and 544.2. These regulations are, for this location, modified
705.415.2.1   by Regulation 705.415.2.1.

In areas intended for livestock, supplementary protective equipotential bonding is required to connect together all exposed-conductive-parts and all extraneous-conductive-parts that can be touched by livestock. Typically, this would include milking parlours and associated areas where cows congregate before and after milking. Areas where livestock retire for the night would also be subject to the requirements for supplementary bonding, as shown in Figure 8.10.

▼ **Figure 8.10** Cattle at rest [photograph courtesy of Wilson Agriculture]

Livestock containment areas pose a number of challenges for the electrical installation designer. Livestock can be panicked by small sensations of current created by differences in potential between exposed-conductive-parts and extraneous-conductive-parts. Keeping exposed-conductive-parts away from livestock is an obvious objective, but this is not always readily achievable. With the proliferation of metal barriers, etc. needed to marshal stock within an area, the requirements of supplementary bonding can be demanding. As can be seen from Figure 8.11, there can be a very large number of metal components, each of which may require supplementary bonding. Although each metal part would require supplementary bonding, it would be impractical and unnecessary to take a conductor to each component part where effective and reliable connection is already made by the construction of the metalwork.

705.544.2 Protective bonding conductors require protection against mechanical damage and corrosion and must be selected to avoid electrolytic effects. Regulation 705.544.2 gives examples of how this may be achieved, namely, by using hot-dip galvanized steel strip with minimum dimensions of 30 mm × 3 mm, or hot-dip galvanized round steel of at least 8 mm diameter, or copper conductors having a minimum cross-sectional area of 4 mm^2.

▼ **Figure 8.11** Metalwork in a typical cowhouse [photograph courtesy of Wilson Agriculture]

Where, for example, the stock areas have concrete slabs (which by their very nature are earthy), it is desirable to lay in a metallic grid within the floor slab to which supplementary bonding conductors are connected. Figure 8.12 shows the recommended connection arrangement. This is, of course, relatively easy at the site construction stage, but highly problematic and costly to retrofit.

▼ **Figure 8.12** A means of supplementary bonding to a metallic grid

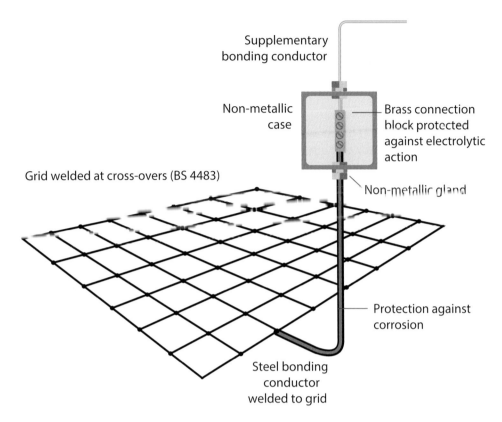

Figure 8.12 illustrates an example of effecting supplementary protective equipotential bonding to a metallic grid. Conductors and connections, which should be accessible for inspection, testing and maintenance, should be selected to withstand all the adverse effects of the external influences likely to be present, such as corrosion and mechanical damage.

526.3 Where connections cannot be made in accessible positions, such as those made to a metal grid within a concrete slab, they are required by Regulation 526.3 to be made by:

▶ welding;
▶ soldering; or
▶ brazing; or
▶ an appropriate compression tool.

At least two connections should be made to the metallic grid, preferably at two or more diagonally extreme points. Additionally, for the metallic grid to be entirely effective, all constituent parts of the grid should be reliably and durably connected together.

414.4.4 Regulation 414.4.4 requires that supplementary protective equipotential bonding is *not* carried out to exposed-conductive-parts of SELV circuits.

Guidance on the csa and the limiting resistance of supplementary bonding conductors is given in clauses 8.3 and 8.4, respectively.

8.10 Conducting locations with restricted movement

Sect 706 Section 706 of BS 7671 addresses the particular requirements for these locations. However, it is important to note that these requirements modify or supplement the general requirements contained elsewhere in BS 7671. In other words, compliance with all the relevant requirements of BS 7671 is essential, not just Section 706.

A *conducting location with restricted movement* is defined as:

Part 2 *A location comprised mainly of metallic or conductive surrounding parts, within which it is likely that a person will come into contact through a substantial portion of their body with the conductive surrounding parts and where the possibility of preventing this contact is limited.*

For example, such a definition may apply to:

▶ a large boiler interior;
▶ a grain silo;
▶ a large diameter metal pipe or large ventilation duct; and
▶ a metal storage tank, the tower or nacelle of a wind turbine.

Figure 8.13 illustrates persons gaining access to two examples of conducting locations permitting only limited, i.e. restricted, bodily movement.

▼ **Figure 8.13** Two examples of conducting locations with restricted movement

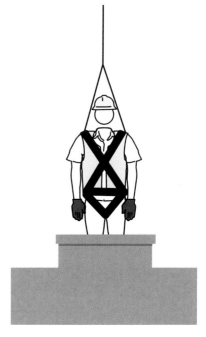

706.410.3.10 Protection against electric shock for fixed equipment can be achieved by one or more of five protective measures (Regulation 706.410.3.10):

▶ SELV;
▶ PELV, where equipotential bonding is provided and the PELV system is earthed;
▶ electrical separation;

▶ use of Class II equipment; and

▶ automatic disconnection with supplementary protective equipotential bonding.

(**Note:** Additional requirements apply to the supply to hand-held tools, items of mobile equipment and handlamps. Refer to Regulation 706.410.3.10.)

415.2
544.2 As indicated above, the protective measure of automatic disconnection requires supplementary protective equipotential bonding to be undertaken in this special location. An additional supplementary bonding conductor is to be provided, meeting the requirements of Regulations 415.2 and 544.2, connecting the exposed-conductive-parts of the fixed electrical equipment within the location and the conductive parts of the location, as shown in Figure 8.14. If PELV is being considered for such a location, similar requirements for equipotential bonding are given in Regulation 706.410.3.10.

706.411.1.2 Where equipment, such as electronic devices with a protective conductor current, requires a facility for functional earthing, Regulation 706.411.1.2 requires that supplementary bonding conductors shall connect together:

▶ all exposed-conductive-parts of equipment;
▶ all extraneous-conductive-parts inside the conducting location with restricted movement; and
▶ the functional earths of any equipment having a protective conductor current.

▼ **Figure 8.14** Conducting location showing the required supplementary bonding conductor

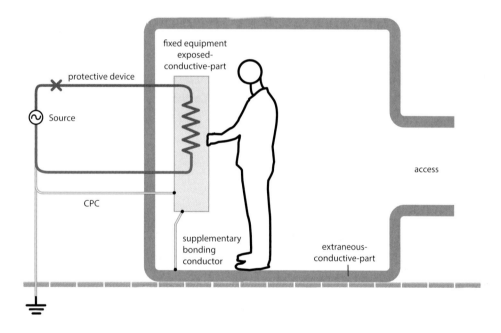

Figure 8.15 shows the connections of supplementary bonding for functional earthing in a conducting location with restricted movement.

▼ **Figure 8.15** Showing the connections of supplementary bonding for functional earthing

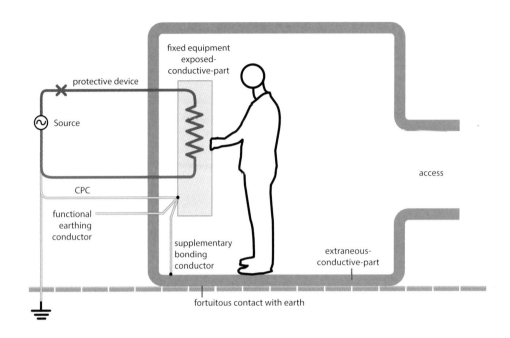

8.10.1 Medical locations

Requirements for supplementary bonding in medical locations are described in section 10.9.7.

Sect 740 ## 8.11 Temporary electrical installations for structures, amusement devices and booths at fairgrounds, amusement parks and circuses

740.415.2.1 In locations intended for livestock, supplementary bonding is to be provided and it must connect all exposed-conductive-parts and extraneous-conductive-parts that can be touched by livestock (Regulation 740.415.2.1). Where a metal grid is laid in the floor, it must be connected to the supplementary bonding in the location. Refer to Figure 8.12.

Extraneous-conductive-parts in, or on, the floor must be connected to the supplementary bonding of the location.

The regulation mentioned above recommends that spaced floors made of prefabricated concrete elements be part of the equipotential bonding. It is important that the supplementary bonding is durably protected against mechanical stresses and corrosion.

8.12 Static convertors

415.2 The installation of a static convertor, such as a UPS, where automatic disconnection
544.2 cannot be achieved for parts on the load side of the static convertor, is required by Regulations 415.2 and 544.2 to be provided with supplementary bonding.

551.4.3.3.1 As required by Regulation 551.4.3.3.1, on the load side of the convertor, supplementary bonding conductors must connect together simultaneously accessible:

▶ extraneous-conductive-parts; and
▶ exposed-conductive-parts.

The resistance of the supplementary bonding conductor is limited to the value given in Equation (8.6):

$$R = \frac{50V}{I_a} \left(\Omega \right) \tag{8.6}$$

where:

R   is the resistance of the supplementary bonding conductor.

I_a is the maximum fault current that can be supplied by the static convertor when the bypass switch is closed.

551.7 Where a static convertor is intended to operate in parallel with a distributor's network, the requirements of Regulation 551.7 also apply.

8.13 Other locations of increased risk

It is recognized that there may be locations of increased risk of electric shock other than those specifically addressed in Part 7 of BS 7671. Examples of such locations could include laundries, where there are washing and drying machines in close proximity and water is present, and commercial kitchens with stainless steel units, where, once again, water is present.

Where, because of the perception of additional risks being likely, the installation designer decides that an installation or location warrants further protective measures, the options available include:

▶ ADS by means of a residual current device having a rated residual operating current ($I_{\Delta n}$) not exceeding 30 mA;
▶ supplementary protective equipotential bonding; and
▶ reduction of maximum fault clearance time.

419.2
544.2 Any supplementary bonding would be required to comply with Regulations 419.2 and 544.2.

8.14 Where automatic disconnection is not feasible

411.3.2.5 Regulation 411.3.2.5 recognises that automatic disconnection within the prescribed time limit with an overcurrent protective device is not always feasible. This might, for example, apply to a circuit with a high current rating where the earth fault loop impedance, Z_s, is too high to cause automatic disconnection within 5 s.

Taking a practical example of a distribution circuit protected by a 200 A BS 88-2 fuse for which the earth fault loop impedance at the most remote end of the circuit is determined to be 0.3 Ω, such a value of Z_s is not sufficiently low to cause automatic disconnection within the limit of 5 s. With such a value, automatic disconnection may take up to about 40 s: clearly, this does not meet the requirements for protection against electric shock.

In such circumstances, the installation designer is afforded two further options:

419.2
544.2

▶ local supplementary protective equipotential bonding in accordance with Regulations 419.2 and 544.2; or

▶ protection by means of an RCD.

Where local supplementary bonding is the chosen option to provide protection against electric shock, consideration should also be given to the thermal effects of the fault current on the cable over the extended period.

Circuit protective conductors

9

9.1 Circuit protective conductors

411.3.1.1
411.5.1
411.6.2
Where automatic disconnection of supply (ADS) is used as the protective measure against electric shock, all exposed-conductive-parts within an electrical installation are required to be earthed to satisfy Regulation 411.3.1.1 for TN systems, Regulation 411.5.1 for TT systems and Regulation 411.6.2 for IT systems. This is achieved by connecting exposed-conductive-parts to the main earthing terminal (MET), via circuit protective conductors (cpcs), and hence to the means of earthing via the earthing conductor. Exposed-conductive-parts may be earthed individually, in groups or collectively (i.e. using a common cpc for a number of circuits).

A cpc therefore provides part of the earth fault loop in the event of an earth fault in the circuit with which it is associated. The main function of a cpc is to carry the earth fault current without damage either to itself or to its surroundings, for example, insulation. In doing so, the earth fault current can be detected by the protective device for the circuit, whether this is an overcurrent protective device, such as a circuit-breaker or fuse, or an RCD. The cpc should be efficient and reliable, in order for the earth fault current to be detected, so that the protective device can provide its function of automatic disconnection within the prescribed time limit.

There are a number of types of cpc in common use, including metal wiring enclosures such as a rigid conduit, trunking, ducting or riser system, as well as the metal covering of cables, such as a sheath or armouring.

There are four instances where a separate cpc is required:

543.2.3
543.7.1.203
543.2.7
543.1

▶ through a flexible or pliable metal conduit;
▶ subject to specified criteria, in a final circuit intended to supply equipment producing a protective conductor current in excess of 10 mA in normal service;
▶ to connect the earthing terminal in an associated box or other enclosure, where the protective conductor is formed by metal conduit, trunking or ducting or the metal sheath or armour of a cable; and
▶ where the armour of a steel wire armoured cable does not meet Table 54.7 requirements (see Appendix B of this publication).

Apart from these instances, there are a number of circumstances in which it is recommended that consideration is given to the provision of a separate cpc if metal conduit, trunking or ducting is used, including in:

▶ industrial kitchens, laundries and other 'wet areas';
▶ locations where chemical attack or corrosion of the metal wiring enclosure or cable sheath may be expected; and

> ▶ circuits exceeding 160 A rating and any location where the integrity of the metal-to-metal joints in the conduit/trunking installation cannot be ensured over the lifetime of the installation. (Where full information concerning correct and reliable earthing is provided by the manufacturer, such systems may be used up to their maximum rating.)

Even where a separate cpc is provided, the metal cable management system is still an exposed-conductive-part containing live conductors and has therefore still to be properly constructed, provide good continuity throughout its run and be earthed.

543.1.2 Where a cpc is used to serve a number of circuits, it should meet the requirement for the most onerous duty. If the csa of the cpc is determined by selection (see Table 3.1 of this Guidance Note), then the csa relating to the larger or largest line conductor of all the circuits that are served by the common cpc has to be used. However, where the method of calculation is used, by means of the adiabatic equation, the most onerous values of the earth fault current and disconnection time should be used. It should be noted that the lowest value of earth fault current may not necessarily produce the lowest I^2t value.

9.2 Cross-sectional area

The csa of a cpc is an important consideration because of the high temperatures that can be generated under earth fault conditions in an inadequately sized conductor.

The temperature rise of a protective conductor under earth fault conditions should therefore be limited to acceptable levels, so that damage to its insulation and sheath, if present, or to adjacent material does not occur.

543.1.1 Regulation 543.1.1 calls for the csa of a protective conductor, other than a protective bonding conductor, to be not less than that:

543.1.3 ▶ calculated in accordance with Regulation 543.1.3; or
543.1.4 ▶ selected in accordance with Regulation 543.1.4.

Most practitioners will agree that the option of selection is often easier to apply than calculation, although it must be emphasised that this can produce a larger csa value for the protective conductor than that obtained by calculation, erring as it does on the safe side.

Regulation 543.1.1 states that if the csa of the line conductors has been determined by consideration of short-circuit current, and if the earth fault current is expected to be less than the short-circuit current, the option of calculation has to be used.

9.2.1 Calculation of csa – general case

543.1.3 Turning first to the calculation option, this is achieved by making use of the adiabatic equation given in Regulation 543.1.3, which was initially presented as Equation (3.1) and is now repeated here for convenience as Equation (9.1):

$$S \geq \frac{\sqrt{(I^2t)}}{k} \, (mm^2) \tag{9.1}$$

where the definitions are the same as those given in Equation (3.1).

The value of S obtained from Equation (9.1) represents the minimum csa value that is required for the cpc. However, there is a minimum value of 2.5 mm² for the csa of a separate protective conductor that is not part of a cable and is not formed by a wiring enclosure or contained within such an enclosure.

In order to use Equation (9.1), values of I and t have to be determined. The prospective earth fault current I can be calculated using Equation (9.2):

$$I = \frac{U_0 \times C}{Z_s} (A)$$

(9.2)

where:

U_0 is the nominal line voltage to Earth.

C   is the voltage factor to take account of voltage variations depending on time and place, changing of transformer taps and other considerations. Whether C_{min} or C_{max} is used will depend on which allows the greater energy let-through (I^2t). For fuses I^2t will usually be greater if C_{min} is used and for circuit-breakers it may be C_{max}.

Z_s is the earth fault loop impedance related to the prescribed disconnection time limit for the circuit.

Note: For a low voltage supply given in accordance with the ESQCR, C_{min} is given the value 0.95, and C_{max} is given the value 1.1.

Table 41.4   To take an example, consider a 230 V final circuit wired in 70 °C PVC cable with an integral copper cpc and protected by a 63 A BS 88-2 fuse. Disconnection is required within 5 s. As given in Table 41.4 of BS 7671, the maximum Z_s is 0.78 Ω, which gives an earth fault current I of 280 A, as shown in Equation (9.3):

$$I = \frac{U_0 \times C_{min}}{Z_s} = \frac{230 \times 0.95}{0.78} = 280\,A$$

(9.3)

Table 54.2   Having determined I for the corresponding value of disconnection time, t, we now
Table 54.3   have to determine a value for k (a constant) by reference to Tables 54.2 to 54.6 as
Table 54.4   appropriate (most of which are replicated in Appendix A). As the cpc is incorporated
Table 54.5   into a cable, Table 54.3 is appropriate, from which we obtain a value for k of 115. We
Table 54.6   can use these values in Equation (9.1), giving a minimum value of the csa of the cpc of 5.44 mm², as shown in Equation (9.4):

$$S = \frac{\sqrt{(I^2 t)}}{k} = \frac{\sqrt{(280^2 \times 5)}}{115} = \frac{626}{115} = 5.44\,mm^2$$

(9.4)

Obviously, as the application of the adiabatic equation produced a non-standard conductor size, a 6 mm² cpc would more than meet the minimum csa obtained by the use of Equation (9.1)

There are situations where the use of separate values of I² and t is impracticable or unreliable, such as:

▶ where the value of I is so high that a corresponding value of t is not obtainable from the time/current characteristic for the protective device;

▶ for short operating times of less than 0.1 s, where current asymmetry is significant (for example, for a protective device installed immediately downstream of the source); and

▶ where the protective device is current-limiting and the prospective earth fault current is of such magnitude that the device will limit the current during fault conditions.

In any of the above circumstances, values of I²t should be obtained from the energy let-through characteristic for the protective device published by the device manufacturer. The I²t value is derived from the manufacturer's characteristic given in a curve showing maximum values of I²t as a function of prospective current under stated operating conditions, or given as a table of values.

9.2.2 Selection of csa – general case

Table 54.7 As previously mentioned, selection can often be an easier way to determine the csa of a cpc. This is done by using Table 54.7 of BS 7671, replicated here as Table 9.1.

▼ **Table 9.1**   Table 54.7 of BS 7671: Minimum csa of protective conductors in relation to the csa of associated line conductors

Csa of line conductor S	Minimum csa of corresponding protective conductor	
	If the protective conductor is of the same material as the line conductor	If the protective conductor is not of the same material as the line conductor
mm²	mm²	mm²
$S \leq 16$	S	$\frac{k_1}{k_2} \times S$
$16 < S \leq 35$	16	$\frac{k_1}{k_2} \times 16$
$S > 35$	$\frac{S}{2}$	$\frac{k_1}{k_2} \times \frac{S}{2}$

Notes:

k_1 is the value of k for the line conductor, selected from Table 43.1 of BS 7671 according to the materials of both conductor and insulation (replicated in Appendix A).

k_2 is the value of k for the protective conductor, selected from Tables 54.2 to 54.6 of BS 7671, as applicable (other than 54.6, replicated in Appendix A).

Using the selection option to determine the csa of a protective conductor is fairly straightforward. For a cpc of the same material as the associated line conductor, reference to Table 9.1 is all that is required, and we find the minimum csa of the cpc in column 2 of the table. This can be summarized as:

- for line conductors up to and including 16 mm², the minimum cpc is the same csa as the line conductor, subject to a minimum of 2.5 mm² or 4 mm² as set out in the following paragraph;
- for line conductors above 16 mm² and up to and including 35 mm², the minimum csa of the cpc is 16 mm²; and
- for line conductors above 35 mm², the minimum csa of the cpc is half that of the line conductor.

543.1.1 If the protective conductor:

- is not an integral part of a cable (such as a 'twin and earth' cable or an armoured cable); or
- is not formed by conduit, ducting or trunking; or
- is not contained in an enclosure formed by a wiring system,

the csa is required to be not less than 2.5 mm² copper equivalent if protected against mechanical damage is provided by, for example, a cable sheath, or 4 mm² if not so protected.

9.2.3 Non-copper cpcs

543.1.1 Where the cpc is not made of copper, an assessment of its copper equivalent (for thermal capacity) is necessary. Equation (9.5) provides the means for this assessment to be made, with the lower limit of csa of a non-copper cpc being given by $S_{m(min)}$, as required by Regulation 543.1.1:

$$S_{m(min)} = S_{c(min)} \times \frac{k_c}{k_m} \ (mm^2) \tag{9.5}$$

where:

$S_{c(min)}$ is the lower limit given in Regulation 543.1.1 for the csa of a copper protective conductor (that is, 2.5 mm² or 4 mm², as appropriate).

k_m is the value of k for a protective conductor of the metal other than copper.

k_c is the value of k for a copper protective conductor.

9.2.4 Evaluation of k

Tables 54.2 to 54.6 Values of k are obtained from Table 43.1 and Tables 54.2 to 54.6 of BS 7671, most of
Table 43.1 which are replicated in Appendix A. Where data is not available from these tables, k may be evaluated by the use of Equation (9.6), which can also be used when the initial temperature differs from that assumed in the tables:

$$k = \sqrt{\left(\frac{Q_c(\beta+20)}{\rho_{20}} \times \ln\left(1+\frac{\theta_f - \theta_i}{\beta + \theta_i}\right)\right)} \tag{9.6}$$

where:

Q_c is the volumetric heat capacity of the conductor material (J/°C mm³).

β is the reciprocal of temperature coefficient of resistivity at 0 °C for the conductor (°C).

ρ_{20} is the electrical resistivity of the conductor material at 20 °C (Ω mm).

θ_i is the initial temperature of the conductor (°C).

θ_f is the final temperature of the conductor (°C).

In   is log to the base 'e'.

Equation (9.6) can be rewritten as Equation (9.7), and the first part of the expression can be seen to be constant for a particular conductor:

$$k = \sqrt{\left(\frac{Q_c(\beta + 20)}{\rho_{20}}\right)} \times \sqrt{\ln\left(\frac{\beta + \theta_f}{\beta + \theta_i}\right)} \tag{9.7}$$

Table 9.2 provides data for use in Equation (9.7).

▼ **Table 9.2** Data for use with (9.7) for evaluating k

Material	β (°C)	Q_c (J/°C mm²)	ρ_{20} (Ω mm)	$\sqrt{\frac{Q_c(\beta+20)}{\rho_{20}}}$
Copper	234.5	3.45×10^{-3}	17.241×10^{-6}	226
Aluminium	228.0	2.50×10^{-3}	28.264×10^{-6}	148
Lead	230.0	1.45×10^{-3}	214.000×10^{-6}	42
Steel	202.0	3.80×10^{-3}	138.000×10^{-6}	78

Table 43.1   As an example of the application of the above data and Equation (9.7), take a copper conductor with 90 °C thermosetting with an initial temperature, θ_i, of 90 °C and a final temperature, θ_f, of 250 °C (from Table 43.1 of BS 7671). Applying Equation (9.7) and using the data given in Table 9.2, we get a value for k of 143, which is in agreement with that given in BS 7671, as confirmed in Equation (9.8):

$$k = 226 \times \sqrt{\ln\left(\frac{\beta + \theta_f}{\beta + \theta_i}\right)} = 226 \times \sqrt{\ln\left(\frac{234.5 + 250}{234.5 + 90}\right)} = 226 \times \sqrt{\ln(1.493)} = 143 \tag{9.8}$$

There are situations for which the designer selects a cable larger than would normally be necessary for the particular load, for example, because of voltage drop considerations. In such a case, the initial temperature would be less than that assumed in Table 43.1 of BS 7671, and, consequently, a higher value for k would apply. For the purposes of illustrating the point, take a similar cable to that previously described and assume an initial temperature, θ_i, of 50 °C. Using the data in Table 9.2 we get a value of 165 for k, as confirmed by Equation (9.9):

$$k = 226 \times \sqrt{\ln\left(\frac{234.5 + 250}{234.5 + 50}\right)} = 226 \times \sqrt{\ln(1.703)} = 165 \tag{9.9}$$

This calculation is only ever likely to be worthwhile where there is a substantial difference in the actual initial temperature to that given in Table 43.1 of BS 7671.

9.3 Armouring

The armouring of a cable, in addition to providing mechanical robustness and protection against impact, may also be used as a cpc. However, as with all cpcs, it is required to meet all the relevant requirements for such conductors. Figure 9.1 shows a typical composition of a four-core steel-wire armoured cable with extruded bedding and oversheath.

▼ **Figure 9.1** Four-core steel-wire armoured cable

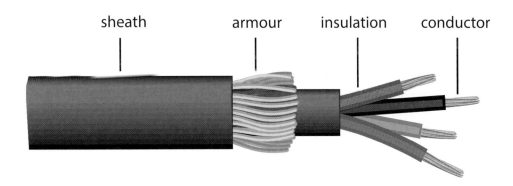

The cable armouring, where used as a cpc, will be subjected to thermal stress during an earth fault due to the earth fault current. As with any other cpc, the csa of the armouring is required to be able to withstand these stresses. In this regard, the csa has to be determined either by calculation or by selection.

9.3.1 Calculation of csa – armoured cable

Table 54.4 For the calculation option, the adiabatic equation has to be used, as given in Equation (9.1). The value for k is obtained from Table 54.4 of BS 7671 (which for ease of reference, is reproduced in Appendix A).

Having obtained the minimum csa using Equation (9.1), this must be checked against the actual csa of the cpc armouring of the intended cable. Data for this comparison for multicore cables with copper and solid aluminium conductors to BS 5467 and BS 6724 is given in Appendix B.

9.3.2 Selection of csa – armoured cable

543.1.1 Selection of the csa of the armouring is much easier than by calculation, and involves the use of the csa of the line conductor, and the k values of both the line conductor and the cpc. However, this method consistently produces a larger minimum csa requirement than by calculation. Selection should not be used where the line conductor(s) are sized only by consideration of short-circuit currents and where the earth fault current is expected to be less than the short-circuit current.

Data for the selection method is given in Appendix B.

9.3.3 Contribution to earth fault loop impedance

Consideration must be given to the contribution that the armouring makes to the overall earth fault loop impedance, Z_s. There may be instances in which reliance on the armouring as a cpc imposes an unacceptable contribution to the overall impedance to the extent that some other provision for a cpc is required.

9.3.4 Armouring inadequate for a cpc

There will be occasions where the actual csa of the armouring of a cable is less than the value of S calculated by use of Equation (9.1). To remedy this, the designer has three options to consider:

514.3.1 **(a)** selecting a cable with an extra internal core and using it as the cpc (note that all the cores are required to be suitably identified – Regulation 514.3.1 refers);
(b) selecting a larger cable with a corresponding larger csa and using the armour as the cpc; or
(c) using a separate green-and-yellow covered copper conductor.

For option (a), it cannot be accurately predicted how the current will divide between what is effectively two parallel conductors (i.e. the core and the armouring), due to the magnetic effect of the armouring. It is therefore important that the additional core is sized as if it alone were to take the earth fault current. In other words, it is not permissible to simply add the csa of the two conductors together.

For option (c), Annex NA of PD CLC/TR 50480 gives information on determining the effective resistance and reactance of the line-protective conductor loop where steel wire armour of a multicore cable is used as a protective conductor. The annex also gives information on checking that the fault current withstand capacity of the external cpc is not exceeded.

9.3.5 Termination of armoured cables

543.3 It is of paramount importance that armoured cables are terminated in a proper manner and in accordance with the manufacturer's installation instructions. BS 7671 demands the preservation of continuity of protective conductors and specifically requires such conductors to be protected against:

- ▶ mechanical damage and vibration;
- ▶ chemical deterioration and corrosion;
- ▶ heating effects; and
- ▶ electrodynamic effects (i.e. mechanical forces generally associated with fault currents).

Even where the armouring is not serving as a cpc, it will still need to be earthed, as it is an exposed-conductive-part. Armouring is therefore always to be protected against the aforementioned detrimental influences and any other stresses that could impair its continuity. Electrical joints, such as those between armouring, cable glands and earthing terminals, are required to be soundly made (mechanically and electrically) and, where necessary, suitably protected.

Regulation 8 of the *Electricity at Work Regulations 1989* places an absolute requirement on protective conductor connections to earth and states:

> *… a conductor shall be regarded as earthed when it is connected to the general mass of earth by conductors of sufficient strength and current-carrying capability to discharge electrical energy to earth.*

Being an absolute requirement, compliance must be achieved regardless of cost or any other consideration.

543.2.2 It may be questionable whether the termination of steel or aluminium wire armouring with glands into metal gland plates, which themselves may only be bolted to the switchgear or controlgear frame, is adequate. Terminations made in this way will always leave some doubt about the effective current-carrying capacity, which needs to be of the order of several kiloamperes, through the metalwork joints, some of which may be electrically impaired by paint or other surface finishes. It is therefore always desirable, and often necessary, to employ 'gland earth tag washers' together with a copper protective conductor with cable lug to bridge the 'gap' between the armouring and the earthing terminal of the item of equipment. Figure 9.2 illustrates an example of a means of terminating an armoured cable. It shows the removable metallic plate at the bottom of the enclosure, together with two retaining screws, through which all the earth fault current would pass were it not for the gland earth tag washer and copper protective conductor arrangement.

▼ **Figure 9.2** An example of a means of terminating an armoured cable

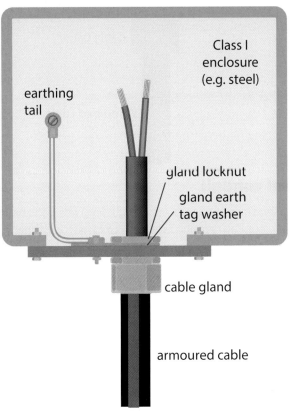

Where an armoured cable is to be terminated into a non-metallic item of equipment, such as an adaptable box, a gland earth tag washer arrangement will be essential where cpc continuity and preservation is to be maintained. The gland earth tag washer arrangement should be clamped between two locknuts to avoid the risk of the degradation of continuity of the connection by creepage of the non-metallic material. Figure 9.3 illustrates such an arrangement

▼ **Figure 9.3** Termination of an armoured cable into a non-metallic item of equipment

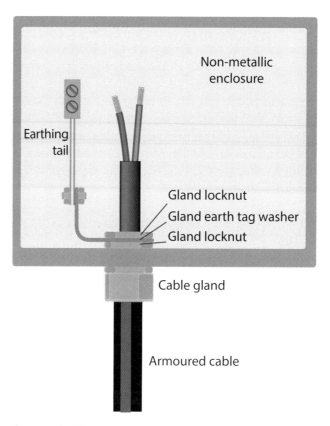

9.4 Steel conduit

543.2.5 Metal conduit (and also steel trunking and ducting) has traditionally been used as a cpc for many years, but has recently fallen out of favour, with many designers adopting a 'belt and braces' approach by using a separate cpc contained within the wiring system. It has to be said that in many, if not most, cases the additional separate cpc is wholly unnecessary. This is because a steel containment system is required to be effectively earthed where the contained cables are not sheathed. Thus, it is important that joints in the containment system are mechanically and electrically sound even where a separate cpc is employed.

A metal conduit often has a large enough csa to allow it to be used as a cpc for the circuit(s) contained within.

543.1.1 As for other forms of cpc, Regulation 543.1.1 requires the csa of the conduit to be either:

543.1.3 ▶ calculated in accordance with Regulation 543.1.3 (essential where the csa of the live conductors has been chosen by considering only the short-circuit current and where the earth fault current is predicted to be less than the short-circuit current); or

543.1.4 ▶ selected in accordance with Regulation 543.1.4.

Using either option, the csa of the cpc has to be sufficient to limit the temperature rise to acceptable limits, thus avoiding any damage to the conductor insulation and, where applicable, the sheaths of the cables and adjacent surroundings.

9.4.1 Calculation of csa – steel conduit

As with the calculation of the csa of any other type of cpc, Equation (9.1) is used.

Fig 3A3(b) To illustrate the calculation option by example, take a 70 °C thermoplastic (PVC) insulated copper cable that is to be installed in a steel conduit, where the conduit is to be used as the only cpc. Fault protection is provided by a 63 A BS 88-2 fuse. Overcurrent protection is provided by the same device. For the purposes of this example, the prospective earth fault current at the furthest point of the circuit is 280 A. As can be seen from Figure 3A3(b) of Appendix 3 of BS 7671 (replicated here as Figure 9.4), for an earth fault current of this magnitude, the overcurrent protective device can be expected to operate in not more than 5 s.

▼ **Figure 9.4** Time/current characteristics for BS 88-2:2013 fuses

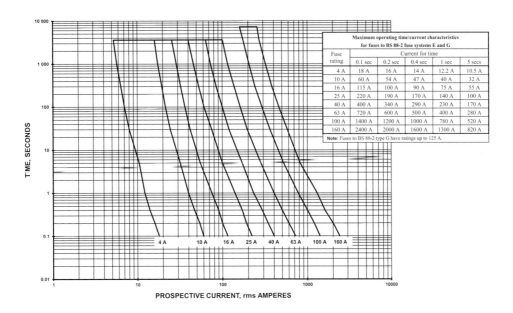

Table 54.5 The other factor required to enable Equation (9.1) to be applied is a value for k, which can be found in Table 54.5 of BS 7671, reproduced in Appendix A, and given as 47. Substituting the data into Equation (9.1), we obtain Equation (9.10), which gives a value of S, the minimum csa for the steel conduit acting as a cpc of:

$$S = \frac{\sqrt{(I^2 t)}}{k} = \frac{\sqrt{(280^2 \times 5)}}{47} = \frac{\sqrt{(392{,}000)}}{47} = \frac{626}{47} = 13.3\,\text{mm}^2 \tag{9.10}$$

To conclude, where the csa is calculated, the minimum csa S of the conduit is required to be not less than 13.3 mm².

Having determined by calculation the minimum csa, this has to be checked against the csa data for steel conduit, given in Table 9.3. It will be noted that even the smallest size steel conduit has a much greater csa than the required 13.3 mm².

▼ **Table 9.3** Csa of heavy-gauge steel conduit to withdrawn standard BS 4568: Part 1:1970

Nominal diameter (mm)	Minimum csa (mm^2)
16	58.8
20	83.1
25	105.5
32	137.3

9.4.2 Selection of csa – steel conduit

Selection of the minimum csa for a cpc provided by a steel conduit involves the following procedure:

Table 43.1
Table 54.5

▶ Utilize the data given in Table 54.7 of BS 7671, together with that in Tables 43.1 and 54.5 of BS 7671 (see Appendix A of this Guidance Note).
▶ It is then necessary to determine the actual csa of the conduit (for example, by reference to the product manufacturer or the relevant British Standard, or from Table 9.3).

Table 54.7
▶ Make certain that the actual csa is not less than the required csa.

Table 9.4 provides data based on the preceding procedure. Taking an example of the application of the data, consider a steel conduit enclosing two circuits:

▶ circuit no. 1: 2.5 mm^2 copper conductors with 90 °C thermosetting insulation; and
▶ circuit no. 2: 6 mm^2 copper conductors with 70 °C thermoplastic (PVC) insulation.

From the table, we can obtain a minimum csa of 6.2 mm^2 and 14.7 mm^2, respectively, for the cpcs for circuit no. 1 and circuit no. 2. Comparing the most onerous csa of 14.7 mm^2 with data in Table 9.3, we see that any conduit size between 16 mm and 32 mm will be adequate from this viewpoint.

▼ **Table 9.4** Minimum csa required for a steel conduit, enclosing copper conductors, determined by using formulae from Table 54.7 of BS 7671

Csa of largest line conductor, S	Table 54.7 formula	Minimum csa of steel conduit (mm^2)		
		Containing 70 °C thermoplastic (PVC) insulated copper cables	Containing 90 °C thermoplastic (PVC) insulated copper cables	Containing 90 °C thermosetting insulated copper cables
		$k_1 = 115$ $k_2 = 47$	$k_1 = 100$ $k_2 = 44$	$k_1 = 143$ $k_2 = 58$
1.5		3.7	3.4	3.7
2.5	$\dfrac{k_1}{k_2} \times S$	6.1	5.7	6.2
4		9.8	9.1	9.9
6		14.7	13.6	14.8
10		24.5	22.7	24.7
16		39.1	36.4	39.4

Csa of largest line conductor, S	Table 54.7 formula	Minimum csa of steel conduit (mm²)		
		Containing 70 °C thermoplastic (PVC) insulated copper cables	Containing 90 °C thermoplastic (PVC) insulated copper cables	Containing 90 °C thermosetting insulated copper cables
		$k_1 = 115$ $k_2 = 47$	$k_1 = 100$ $k_2 = 44$	$k_1 = 143$ $k_2 = 58$
25	$\frac{k_1}{k_2} \times 16$	39.1	36.4	39.4
35		39.1	36.4	39.4
50	$\frac{k_1}{k_2} \times \frac{S}{2}$	61.2	56.8	61.6
70		85.6	79.5	86.3

9.4.3 Maintaining protective conductor continuity

543.3.6 As previously mentioned, irrespective of whether the steel conduit is used as a cpc or not, the conduit itself is required to be earthed. It is therefore essential that all joints in the conduit are mechanically robust and electrically sound. This requires that the screwed joints in the system, the conduit couplers and the conduit boxes are clean before being tightened. Similarly, bushes and couplers used for terminating the steel conduit into accessory boxes and the like are required to be tightened by the use of the correct tool. The conduit system is required to be selected with reference to environmental conditions. Black enamelled and galvanized steel conduit are available. respectively, for use in dry and other conditions. For very arduous duties, stainless steel conduit is also available.

9.5   Steel trunking and ducting

The csa of a cpc formed by steel trunking or ducting is determined in the same manner as for steel conduit, either by calculation using Equation (9.1) or by selection using data given in Table 9.4. Trunking used in a new installation should conform to BS EN 50085 and there are no minimum csa values in that standard. For older installations, Table 9.5 gives csa values for steel trunking to BS 4678 Part 1:1971, which is now withdrawn.

▼ **Table 9.5** Csa of steel trunking to withdrawn standard BS 4678:Part 1:1971

Nominal size (mm × mm)	Minimum csa (mm²) without lid
38 × 38	103
50 × 38	113
50 × 50	135
75 × 50	189
75 × 75	243
100 × 50	216
100 × 75	270
100 × 100	324
150 × 50	270

Nominal size (mm × mm)	Minimum csa (mm²) without lid
150 × 75	324
150 × 100	378
150 × 150	567

9.5.1 Maintaining protective conductor continuity

As with steel conduits, steel trunking will constitute an exposed-conductive-part even where not used as a cpc. It is therefore of paramount importance that sections of trunking, trunking bends and trunking terminations are mechanically robust and electrically sound. Where specified by the manufacturer, continuity links should be used to link two sections together, or a section and a bend, although a suitable additional conductor may be required between sections of a metallic wiring system and between such a system and the equipment enclosures to which it connects. Trunking bends such as angled bends (for example, 45 °, 90 ° and 135 °) supplied by the manufacturer should always be used.

9.6 Other metal enclosures

Where a cpc is made up in part of a metal enclosure or frame of low voltage switchgear or controlgear, it is important that effective electrical continuity and fault current carrying capability is maintained, either by construction or by suitable connection.

543.2.2 BS 7671 requires that the csa of the enclosure or frame must be not less than that calculated by Equation (9.1) or must be selected by reference to Table 54.7 of BS 7671. Alternatively, this may be verified by a test procedure in accordance with the appropriate part of the BS EN 61439 series.

If there is doubt as to whether the above requirements are met, a gland earth tag washer and copper protective conductor arrangement should be provided, as shown in Figure 9.2. For Class I equipment, the additional gland locknut shown in Figure 9.3 is not essential if the gland earth tag washer can be fitted directly in contact with a cleaned metallic surface of the enclosure. For non-metallic enclosures and Class II equipment, the additional locknut should always be used.

9.7 Terminations in accessories

As conductors for safety, the proper termination of cpcs is essential in every instance. This is no less so where the cpc is terminated into an accessory. The effectiveness of the termination is reliant upon the good workmanship and diligence of the electrician.

543.2.7 For accessories where the cpc is provided by a metal conduit, or by trunking or ducting, or the metal sheath or armour of a cable, Regulation 543.2.7 of BS 7671 requires an earthing tail to be fitted that connects the earthing terminal of the accessory to the earthing terminal incorporated in the associated back box. Figure 9.5 shows this earthing tail fitted so as to link an accessory to a dado-height metallic trunking system and a similar accessory to a flush conduit system back box.

▼ **Figure 9.5** 'Earthing tail' to an accessory

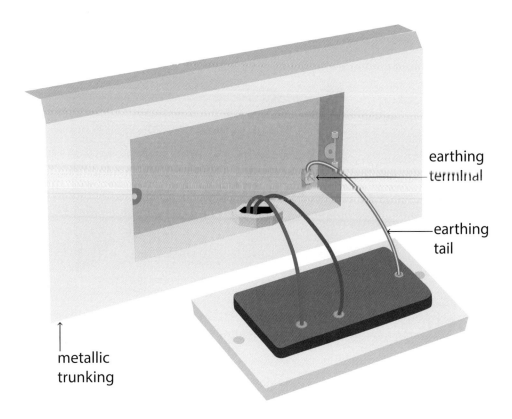

earthing terminal

earthing tail

metallic trunking

In the example shown in Figure 9.5, the accessory box is earthed to the metal conduit, trunking, ducting (or metal sheath or armour of a cable, if this were to be the wiring system) forming the cpc. The purpose of the earthing tail is not to earth the back box, but to earth the accessory by connecting it to the back box and hence to the cpc.

A metal back box for a surface-mounted accessory is an exposed-conductive-part, and a metal back box for a flush-mounted accessory is deemed to be an exposed-conductive-part. Therefore, such back boxes, like all exposed-conductive-parts, are required to be earthed.

Determining how to achieve this is often the subject of much debate; most choose to rely upon the use of an earthing tail between the accessory and the associated back box. Whether such an arrangement is required largely depends upon whether one or both lugs of the back box are adjustable and upon the earth strap and eyelet arrangement of the accessory, all combinations of which are detailed as follows.

For flush metal back boxes with two fixed lugs, as shown in Figure 9.6, the box can be considered adequately earthed through the earthing straps and eyelets of the accessory and the fixing lugs on the box. Consequently, an earthing tail in these circumstances is not essential, although some installers still like to provide one, as it is considered good practice.

▼ **Figure 9.6** Flush metal box with two fixed accessory-fixing lugs

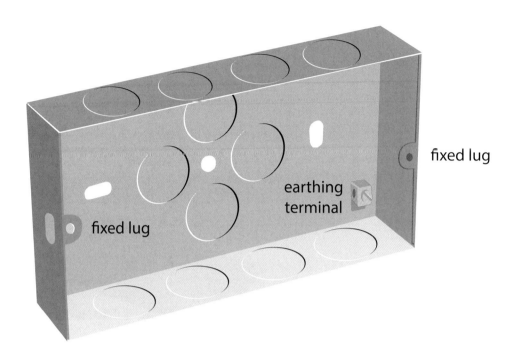

For flush metal back boxes with two adjustable accessory-fixing lugs, as shown in Figure 9.7, an earthing tail is essential because these lugs cannot be relied upon for continuation of the cpc to earth the accessory. Experience has shown that the lugs are subject to corrosion and can present a high-resistance connection that will impede the earth fault current and may prevent the device for automatic disconnection from operating.

▼ **Figure 9.7** Flush metal box with two adjustable accessory-fixing lugs

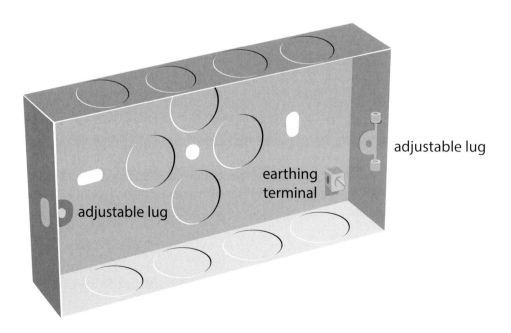

For flush metal back boxes with one fixed lug and one adjustable accessory-fixing lug, as shown in Figure 9.8, it is always desirable to provide an earthing tail. However, because some accessories have only one earthing strap and eyelet, it is essential that the earthing eyelet is located at the fixed lug position, otherwise continuation of the cpc will not be maintained and an earthing tail will be required to be provided.

▼ **Figure 9.8** Flush metal box with one adjustable and one fixed accessory-fixing lug

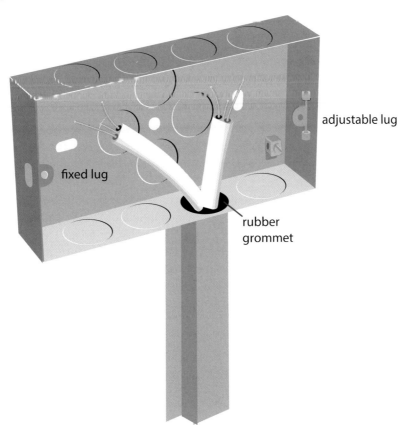

9.8   Cpc for protective and functional purposes

Confusion sometimes arises over the difference between protective earthing and functional earthing. As previously mentioned, protective earthing, as the name suggests, is provided for the protection of people, livestock and property. A functional earth, however, is only provided to enable equipment to operate correctly. At no time does the functional earthing offer protection to either the user or the equipment. Examples of the purposes to which functional earthing is put include:

▶ to provide a 0 V reference point;
▶ to enable an electromagnetic screen to be effective; and
▶ to provide a signalling path for some types of communications equipment.

The most common use of functional earthing is for telecommunications purposes. It is permissible for the functional earthing conductor to be terminated at the electrical installation MET. The wiring used will normally be of copper with a csa not less than 1.5 mm². It should be identified by the colour cream, as recommended in BS 6701:2016 and Table 51 of BS 7671. Additionally, the functional earthing conductor should have a label (or, alternatively, embossed sheath) reading 'Telecoms Functional Earth' where terminated at the MET. Electrical installation practitioners who come across such functional earthing, which is particularly common in commercial buildings, should not interfere with these connections.

Functional earthing may also be required for other equipment and should be identified by the colour cream. The connection to earth should again be made to the MET and be clearly labelled as to its purpose.

543.5.1
543.4.2 Where earthing arrangements are installed for combined protective and functional purposes, the requirements for protective measures are required always to take precedence. A particular application of this is the use of a combined protective and neutral (PEN) conductor in an installation forming part of a TN-C system. The detailed requirements of BS 7671 for such an installation are tightly drawn. Additionally, an exemption to the ESQCR has to be obtained by consumers for adoption of a TN-C system, which is fairly rare.

9.9 Significant protective conductor currents

543.7 Most items of current-using equipment exhibit some current flow in their cpc and generally this is of a magnitude that does not cause concern. However, some items are more prone to show evidence of protective conductor currents which cannot be ignored and in these cases provision has to be made to employ cpcs that have a higher integrity. This is because if a cpc becomes discontinuous for whatever reason, such protective conductor currents, when interrupted, will cause a potential difference between items connected to either end of the conductor, which can present a risk of electric shock. Regulation 543.7 *Earthing requirements for the installation of equipment having high protective conductor currents* addresses the particular requirements for the cpcs for such equipment.

Figure 9.9 shows one example of an item of information technology (IT) equipment that is likely to produce some measure of protective conductor current.

▼ **Figure 9.9** Item of IT equipment causing a protective conductor current

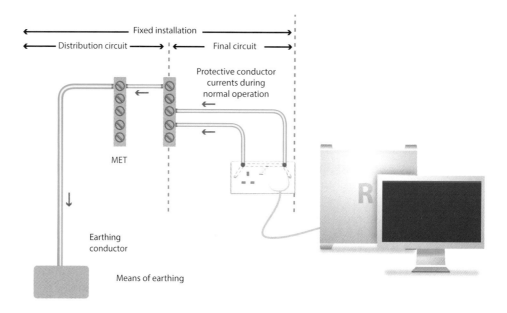

In Figure 9.9, the computer is equipped with filters or suppressors that generate a protective conductor current. Other IT equipment, telecommunications equipment and equipment such as some types of fluorescent luminaires are also known to cause such currents to flow.

Filters are often provided in such equipment to save downstream equipment from harm that might otherwise be caused by transient or switching overvoltages originating from other parts of the fixed installation or from the supply. Where switched mode power supplies are involved, suppressors may serve the purpose of preventing high-frequency noise being transferred into the power supply. Filters and suppressors commonly use capacitive and inductive components between the live conductors (line and neutral conductors) and the cpc.

543.5.1 As mentioned in Section 9.8, and as required by Regulation 543.5.1, where the cpc is provided for combined fault protection and functional purposes, the protective requirements have to take precedence.

In order to improve light output, reduce controlgear losses, provide silent operation and reduce perceptible flicker, many fluorescent luminaires operate at high frequency (for example, 30 kHz). However, to prevent any high-frequency noise being impressed on the supply, these luminaires are often fitted with suppressors which, by design, produce protective conductor currents. Sufficient quantities of luminaires can produce protective conductor current magnitudes that are significant and of concern.

The risk associated with final circuits with significant protective conductor currents results from discontinuity of the protective conductor. For example, in installations where there are significant protective conductor currents, serious risks from electric shock can arise from accessible conductive parts connected to protective conductors that are not connected to the MET. This risk is extended to all the items of equipment on the particular circuit, whether or not they individually have significant protective conductor currents. The more equipment that is connected to a circuit, the wider the risk is spread.

543.7.1.203 IEC publication PD IEC/TS 60479-1:2005+A1:2016: *Effects of current on human beings and livestock* advises on the physiological effects of current passing through the human body (see Figure 9.10 and Table 9.6). At body currents of 10 mA and less, the duration of the current that may result in undesirable physiological effects exceeds 2000 s, which is over 30 minutes. Regulation 543.7.1.203 requires protective measures to be taken for circuits with protective currents exceeding this value (10 mA).

▼ **Figure 9.10** Conventional time/current zones of effects of AC currents (15 to 100 Hz) on persons for a current path corresponding to left hand to feet, from DD IEC/TS 60479-1:2005 Figure 20 (for explanation, see Table 9.6.)

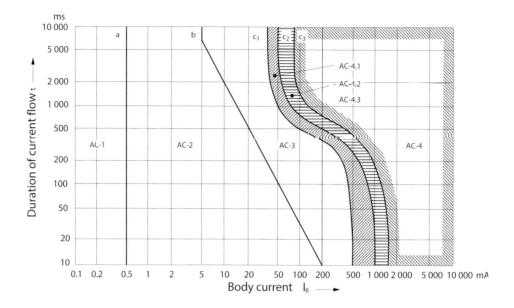

When equipment has a protective conductor current of 10 mA, the impedance that allows this at 230 V is 23 000 Ω. The body plus footwear impedance is usually about 2000 Ω (in dry conditions). Consequently, the body impedance does not effectively limit the touch current. If a person touches the exposed-conductive-parts of this equipment when the protective conductor is disconnected, the current conducted through the body is 230 V/(23 000 + 2000 Ω), that is, 9.2 mA, a minimal reduction.

▼ **Table 9.6** Physiological effects of current on the human body (Table 11 of PD IEC/TS 60479-1:2005)

Zone designation	Zone limits	Physiological effects
AC-1	Up to 0.5 mA (line a)	Perception possible but usually no 'startled' reaction.
AC-2	0.5 mA up to line b	Perception and involuntary muscular contractions likely but usually no harmful electrical physiological effects.
AC-3	Line b and above	Strong involuntary muscular contractions. Difficulty in breathing. Reversible disturbances of heart function. Immobilisation may occur. Effects increasing with current magnitude. Usually no organic damage to be expected.
AC-4	Above curve c1	Patho-physiological effects may occur such as cardiac arrest, breathing arrest, and burns or other cellular damage. Probability of ventricular fibrillation increasing with current magnitude and time.

Zone designation	Zone limits	Physiological effects
AC-4.1	Between c1 and c2	Probability of ventricular fibrillation increasing up to about 5 %.
AC-4.2	Between c2 and c3	Probability of ventricular fibrillation up to about 50 %.
AC-4.3	Beyond curve c3	Probability of ventricular fibrillation above 50 %.

Note: For durations of current flow below 200 ms, ventricular fibrillation is only initiated within the vulnerable period if the relevant thresholds are surpassed. As regards ventricular fibrillation, Figure 9.10 relates to the effects of current that flows in the path left hand to feet. For other current paths, the heart current factor given in PD IEC/TS 60479-1:2005+A1:2016 will have to be considered.

543.7 From Figure 9.10 and Table 9.6, protective conductor currents exceeding 10 mA can have harmful effects, and precautions need to be taken. The requirements of Regulation 543.7 of BS 7671 are intended to increase the reliability of the connection of protective conductors to equipment and to earth, when the cpc current exceeds 10 mA. This normally requires duplication of the protective conductor or an increase in its csa. The increased csa is not to allow for thermal effects of the protective conductor currents, which are insignificant, but to provide for greater mechanical robustness and thereby a more reliable connection to earth. Duplication of a protective conductor, each with independent terminations, is likely to be more effective than an increase in csa.

9.9.1 Equipment

BS EN 60950 series, the standard for the safety of IT equipment, including electrical business equipment, requires equipment with a protective conductor current exceeding 3.5 mA to have an internal protective conductor csa not less than 1 mm^2 and also requires a label bearing the wording given in Figure 9.11 or Figure 9.12, or similar wording, fixed adjacent to the equipment primary power connection.

▼ **Figure 9.11** Equipment warning notice

> **HIGH LEAKAGE CURRENT**
> **Earth connection essential**
> **before connecting the supply**

▼ **Figure 9.12** Equipment warning notice

> **WARNING HIGH TOUCH CURRENT**
> **Earth connection essential**
> **before connecting the supply**

543.7.1.201 Regulation 543.7.1.201 requires a single item of equipment having a protective conductor current exceeding 3.5 mA but not exceeding 10 mA to be either permanently connected to the fixed wiring without the use of a plug and socket-outlet or to be connected by a plug and socket-outlet complying with BS EN 60309-2, as illustrated in Figure 9.13.

▼ **Figure 9.13** BS EN 60309-2 plug [illustration courtesy of Legrand Electric Limited]

543.7.1.202 If the protective conductor current exceeds 10 mA, the requirements of Regulations
543.7.1.203 543.7.1.202 and 543.7.1.203 for a high-integrity protective earth connection should be met.

9.9.2 Labelling at distribution boards

Distribution boards supplying circuits with significant protective conductor currents are required to be labelled accordingly, so that persons working on the distribution boards can maintain the protective precautions already in place.

543.7.1.205 Where a circuit has, or is likely to have, a significant protective conductor current, the protective conductor connection arrangements at the distribution board will be affected by, for example, the sequence of connections. Additional information indicating those circuits that have a significant protective conductor current is required to be made available in the form of a label and should be positioned so as to be visible to a person modifying or extending the circuit, as required by Regulation 543.7.1.205.

9.9.3 Ring final circuits

543.7.2.201 Ring final circuits provide duplicate protective conductors, and if the ends of each protective conductor are separately terminated at the distribution board and at the socket-outlets, the requirements of Regulation 543.7.2.201 will be met, as shown in Figure 9.14. Socket-outlets and other accessories are available with two earth terminals for this purpose, both of which should be used to uphold this increased integrity of the cpc.

▼ **Figure 9.14** Ring final circuit supplying socket-outlets (total protective conductor current exceeding 10 mA)

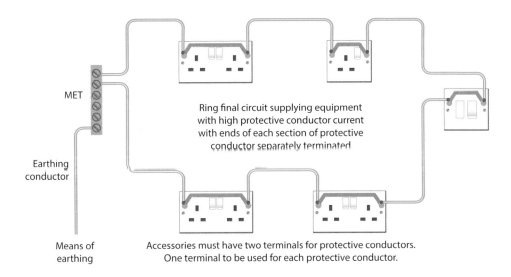

MET

Earthing conductor

Ring final circuit supplying equipment with high protective conductor current with ends of each section of protective conductor separately terminated

Means of earthing

Accessories must have two terminals for protective conductors. One terminal to be used for each protective conductor.

9.9.4 Radial final circuits

543.7.2.201 Radial final circuits supplying socket-outlets that are known or reasonably expected to have a total protective conductor current in excess of 10 mA in normal service are required to have a high-integrity protective conductor connection. Whilst several options are given in Regulation 543.7.2.201, this can often be most effectively provided by a separate duplicate protective conductor connecting the last socket-outlet directly back to the distribution board, as shown in Figure 9.15. This will provide a duplicate connection for each socket-outlet on the circuit. The following requirements apply:

▶ all socket-outlets are required to have two protective conductor terminals, one for each protective conductor; and
▶ the duplicate protective conductors are required to be separately connected at the distribution board.

To reduce interference effects, the duplicate protective conductor should be run in close proximity to the other conductors of the circuit.

▼ **Figure 9.15** Radial circuit supplying socket-outlets (total protective conductor current exceeding 10 mA), with duplicate protective conductor

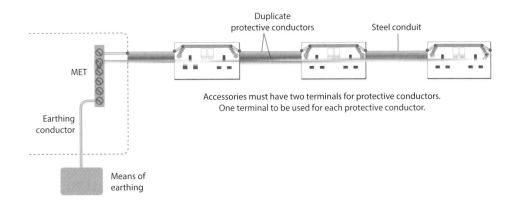

Duplicate protective conductors

Steel conduit

MET

Earthing conductor

Means of earthing

Accessories must have two terminals for protective conductors. One terminal to be used for each protective conductor.

9.9.5   Busbar systems

543.7.1.203   Busbar systems are often adopted by designers for supplying IT equipment (Figure 9.16). These may be radial 30 A or 32 A busbars with tee-offs to individual socket-outlets. The main protective earth (PE) busbar will need to meet one or more of the requirements of Regulation 543.7.1.203, having a protective conductor with a csa not less than 10 mm² or duplicate protective conductors each with a csa sufficient to meet the requirements of Section 543.

▼ **Figure 9.16** Spurs from a 30 A or 32 A busbar

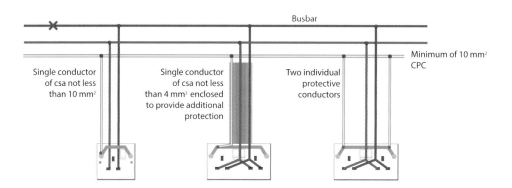

The socket-outlets are often connected as spurs off such a busbar system. If the protective conductor current is expected to be less than 3.5 mA, there is no need for duplication of the protective conductor on the spur from the radial busbar to the socket-outlet. However, if the protective conductor current is likely to exceed 10 mA, then a single copper protective conductor having a csa of not less than 4 mm² and enclosed to provide additional mechanical stability may be used.

It is necessary for the designer to confirm that the disconnection time of the 30 A or 32 A device feeding the busbar is sufficiently fast to provide adiabatic fault protection to the spurred conductors. This almost certainly will be so for, say, 2.5 mm² conductors where a disconnection within 0.4 s is required, but this should be confirmed. Overload protection is not required on the socket-outlet spur from the busbar provided that the spur cable rating exceeds 13 A, as the outlet is a standard 13 A outlet and only plugs fitted with a maximum 13 A fuse can be connected. Hence, overload of the conductor cannot occur, and the 30 A or 32 A device can provide adequate fault protection.

9.9.6   Connection of an item of equipment (protective conductor current exceeding 10 mA)

The connection is required to be made by one of three methods:

▶ a permanent connection to the fixed wiring of the installation; or
▶ a plug and socket-outlet complying with BS EN 60309-2; or
▶ by employing an earth monitoring system conforming to BS 6444.

543.7.1.203   Equipment with a protective conductor current exceeding 10 mA should preferably be permanently connected to the fixed wiring of the installation, with the protective conductor meeting the high-integrity requirements detailed below. It is permitted for the final connection between the item of equipment and the wiring of the installation to be made by means of a flexible cable. The high-integrity requirements of Regulation 543.7.1.203 give five options:

(a) A single protective conductor having a csa of not less than 10 mm².

(b) A single copper protective conductor having a csa of not less than 4 mm², the protective conductor being enclosed in, say, a flexible conduit to provide additional protection against mechanical damage.

543.2.5 **(c)** Two individual protective conductors, each one complying with the requirements of Section 543 of BS 7671, which covers the csa and types of permitted protective conductors. It is permitted for the two protective conductors to be of different types, such as a metallic conduit and a protective conductor enclosed within the same conduit. The two individual protective conductors may also be incorporated in the same multicore cable, provided that the total csa of all the conductors, including the live conductors, is not less than 10 mm². One of the protective conductors may be formed by the metallic sheath, armour or wire braid screen incorporated in the cable (as shown in Figure 9.17), provided that it meets the requirements of Regulation 543.2.5.

(d) An earth monitoring system conforming to BS 4444 configured to automatically disconnect the supply to the equipment in the event of a continuity fault in the protective conductor.

543.7.1.203 **(e)** A double-wound transformer or equivalent unit, such a motor-alternator in which the input and output circuits are electrically separate. In this case, the cpc is required to connect the exposed-conductive-parts of the equipment and a point of the secondary winding of the transformer or equivalent device to the MET of the installation, and the protective conductor between the equipment and the transformer is required to comply with the requirements of Regulation 543.7.1.203. It is important to note that the limits on the earth fault loop impedance associated with fault protection should be met, taking into account the added impedance represented by the transfer function of the transformer.

Equipment with a protective conductor current exceeding 10 mA may be connected to the fixed wiring of an installation by a plug and socket-outlet complying with BS EN 60309-2 (as for equipment with a protective conductor current between 3.5 mA and 10 mA), provided that one of two additional requirements relating to the csa of the protective conductor is met. The first requirement is that the csa of the protective conductor of the associated flexible cable is not less than 2.5 mm² for plugs rated at 16 A, and not less than 4 mm² for plugs rated above 16 A. Alternatively, the protective conductor of the associated flexible cable should have a csa that is not less than that of the line conductor.

▼ **Figure 9.17** One duplicate protective conductor formed by the armour

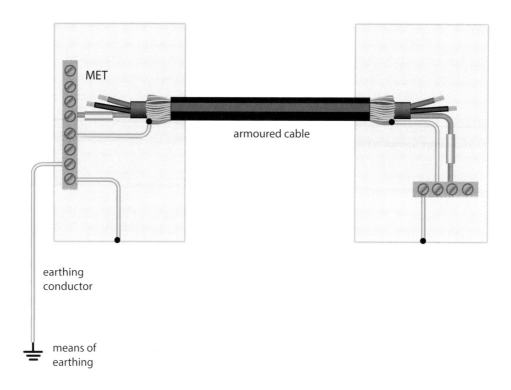

9.9.7 TT systems

If items of equipment having a protective conductor current exceeding 3.5 mA in normal service are to be supplied from an installation forming part of a TT system, the product of the total protective conductor current (in amperes) and the resistance of the installation earth electrodes (in ohms) should not exceed 50, as shown in Equation (9.11):

$$I_{pc} \times R_A \leq 50\,V \tag{9.11}$$

where:

I_{pc} is the protective conductor current.

R_A is the sum of the resistances in ohms of the earth electrode and protective conductor for the exposed-conductive-parts.

543.7.1.203   If this condition cannot be achieved, the equipment should be supplied through a double-wound transformer or equivalent (see Regulation 543.7.1.203). This is to limit the voltage, a consequence of the protective conductor current between earth and any equipment connected to the earth electrode, to 50 V.

9.9.8 RCDs

A protective conductor current in a circuit will be detected by any RCD in that circuit, which could result in the device operating during normal operation or during switching surges.

531.3.2   An RCD should be so selected and the electrical circuits so subdivided that any protective conductor current expected to occur during normal operation of the connected load(s) will be unlikely to cause unwanted tripping of the device (Regulation 531.3.2 refers).

To avoid unwanted tripping, the designer should take care that the anticipated total protective conductor current does not exceed 30 % of the rated residual operating current, $I_{\Delta n}$, of the RCD. Therefore, it may be necessary to provide a number of separate RCD-protected circuits, each designed to supply a limited number of items appropriate to their protective conductor current.

9.10 Earth monitoring

Earth proving and earth monitoring systems conforming to BS 4444 are a recognised means of verifying on a continuous basis the integrity of a protective conductor. In certain installations it may be necessary to know that the integrity of a protective conductor has not been adversely affected, especially where the protective conductor may be particularly vulnerable (for example, in trailing cables, which can suffer mechanical damage or excessive strain).

During the lifetime of an installation, the continuity of a protective conductor may deteriorate due to corrosion, impact or vibration. It may also suffer a loose connection, sustain deterioration or damage or there may be an unintentional disconnection. A protective conductor so affected may then no longer provide an adequate return path for the fault current in the event of an earth fault. Consequently, the effectiveness of fault protective measures that rely on a sound earth return path may be impaired or even rendered completely ineffective.

Application of this earth proving and protective conductor monitoring might include:

▶ construction sites where flexible cables incorporating protective conductors supply items of a movable plant;
▶ any fixed installations where reassurance of the integrity of the protective conductor is required; and
▶ the supply to equipment with significant protective conductor currents.

Both earth monitoring and protective conductor proving systems require a return path for the proving or monitoring current. The return path generally consists of an insulated conductor, known as the pilot conductor. Such systems are the subject of BS 4444.

BS 4444:1989 *Guide to electrical earth monitoring and protective conductor proving* provides definitions:

▶ Earth proving system: A system providing a means of maintaining a high degree of confidence in the continuity of a protective conductor within an installation.
▶ Earth monitoring system: A system for providing a means of maintaining a high degree of confidence in the measured impedance of the protective conductor forming part of the earthing arrangements of an electrical system.

A protective conductor proving system verifies the continuity of a selected protective conductor, whereas an earth monitoring system monitors the impedance of the selected protective conductor.

A typical earth proving unit contains an extra-low voltage supply denoted by 'V'; a sensing device denoted by 'S'; an alarm denoted by 'A'; and the contacts of a relay or a contactor denoted by 'D', as shown in Figure 9.18. The low/extra-low voltage supply provides a current that circulates in the loop circuit consisting of the protective conductor, a part of the metal casing of the protected Class I equipment, the pilot conductor and the sensing device, S. The sensing device S is energized when sufficient current flows in the loop circuit. Contact D then closes, connecting the load circuit to the supply. If the magnitude of the current falls below a prescribed value, the relay de-energizes, disconnecting the load and activating the alarm. In the simple earth proving system shown in the figure, the voltage supply is a transformer; the sensing device, a relay; and the alarm, a warning lamp or audible alarm.

Reference should be made to BS 4444 for further guidance on this aspect of providing high-integrity cpcs.

▼ **Figure 9.18** Simplified circuit for earth proving unit

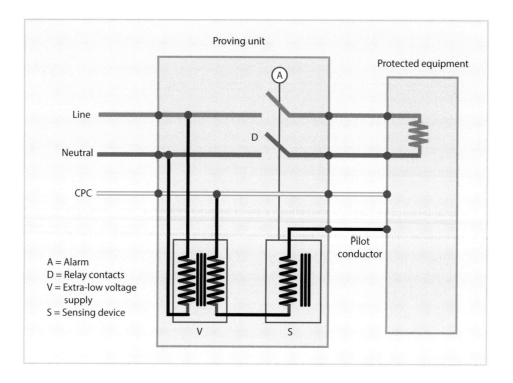

9.11 Proving continuity

GN3 Every protective conductor, including the earthing conductor and protective and supplementary bonding conductors, as well as cpcs, should be inspected and tested to verify that the conductors are electrically sound and correctly connected. IET publication Guidance Note 3: *Inspection & Testing* gives guidance.

Particular issues of earthing and bonding

10

10.1 Clean earths

A 'clean earth' is a particular form of functional earthing that provides a connection to earth, which is necessary for the proper functioning of electrical equipment. A clean earth is more accurately described as a low-noise earth, in which the level of conducted or induced interference from external sources does not produce an unacceptable incidence of malfunction in the data-processing or similar equipment to which it is connected.

Where IT equipment is under consideration, functional earthing and a clean earth or low-noise earth should also be considered.

444.1 Regulation 444.1 of BS 7671 refers designers to BS EN 50310:2010 *Application of equipotential bonding and earthing in buildings with information technology equipment.*

The protective conductors in a building are subject to transient voltages relative to the general mass of Earth. These transient voltages are termed 'earth noise' and are often caused by load switching. They may also be generated by the charging of an equipment frame via the stray capacitance from a low voltage circuit, or mains-borne transients may be coupled into the earth conductor or frame from supply conductors.

As earth noise can cause malfunction, manufacturers of large computer systems usually make specific recommendations for the provision of a 'clean' mains supply and a 'clean' earth. The equipment manufacturer's guidance should be taken into account for such installations.

A dedicated earthing conductor may be used for a computer system, provided that:

(a) all accessible exposed-conductive-parts of the computer system are earthed, the computer system being treated as an 'installation' where applicable; and

542.4.1 **(b)** the MET or bar of the computer system (installation) is connected directly to the building MET by a protective conductor; extraneous-conductive-parts within reach of the computer systems are bonded, but not via the protective conductor referred to in (a) above (see Regulation 542.4.1).

Supplementary bonding between extraneous-conductive-parts and the accessible conductive parts of the computer system is not necessary.

Table 51 Functional earthing conductors used when providing a clean earth are required to be identified by the colour cream, according to Table 51 of BS 7671.

GN1 See also section 10.12 and IET Guidance Note 1: *Selection & Erection.*

10.2 PME for caravan parks

708.411.4   Regulation 9(4) of the ESQCR prohibits the connection of any metalwork in a caravan (or boat) to the combined protective and neutral conductor of the public electricity network. Consequently, the earthing contacts and extraneous-conductive-parts in the form of, for example, metal cases of caravan pitch socket-outlets, should not be connected to a PME earthing terminal, where one is made available by the electricity distributor at the site supply intake position.

708.55.1.2   Figure 10.1 shows a typical caravan park distribution layout where the park supply distribution is connected to the incoming supply with PME. The park installation consists of distribution circuits to the two pillars, which then feed the individual caravans.

708.415.1   The park installation is therefore part of a TN-C-S system with PME, which is permitted by BS 7671. However, Regulation 708.415.1 requires that each socket-outlet on the pillar is protected individually by a residual current device having a rated residual current, $I_{\Delta n}$, of not more than 30 mA that disconnects all live conductors, including the neutral conductor. Regulation 708.553.1.14 states that the protective conductor must not be connected to the PME terminal.

▼ **Figure 10.1** A typical caravan park distribution arrangement

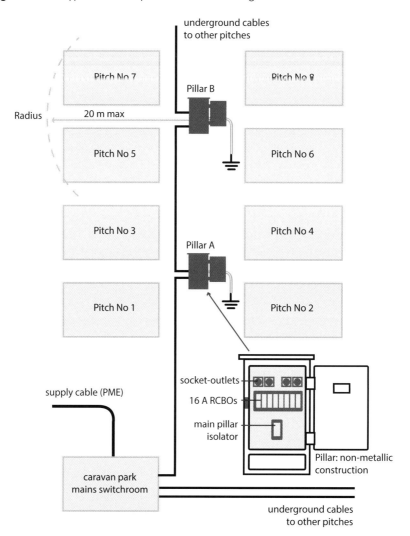

As can be seen from Figure 10.1, the protective conductor of each socket-outlet circuit is connected to an installation earth electrode and the requirements for a TT system should be met.

The demarcation of the park pitch TT system from the electricity distributor's PME earthing terminal may be made at one of a number of places, such as:

▶ at the pitch supply position, as shown in Figure 10.2
▶ at the consumer's distribution position, as shown in Figure 10.3.

▼ **Figure 10.2** Demarcation of TN-C-S/TT systems at a park pillar

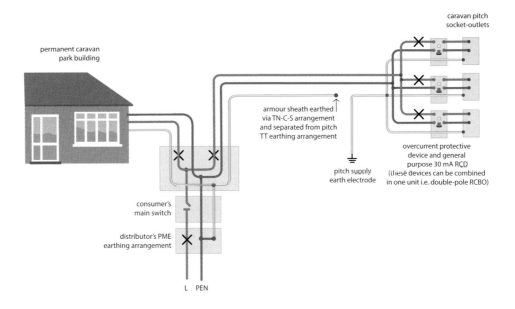

▼ **Figure 10.3** Demarcation of TN-C-S/TT systems at the consumer's intake position

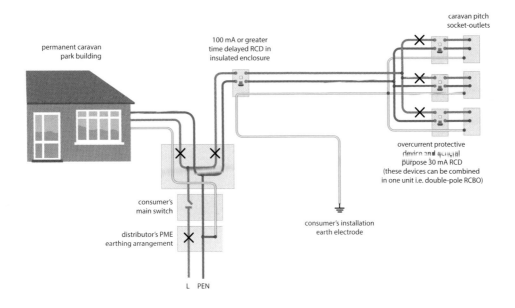

10.3 Exterior semi-concealed gas meters

The installation of outdoor semi-concealed gas meter boxes has been common practice for a number of years.

544.1.2 Regulation 544.1.2 requires the connection of a main protective bonding conductor to any extraneous-conductive-parts to be made as near as practicable to the point of entry of that part into the premises. Where there is a meter, isolation point or union, the connection should be made to the consumer's hard metal pipework and before any branch pipework. Where practicable, the connection should be made within 600 mm of the meter outlet union or at the point of entry to the building where the meter is external. Although it is generally preferable for a suitable bonding connection to be made inside the premises, this may not always be practicable.

In many situations it is not possible to connect the main protective bonding conductor at the point that the gas installation pipe enters the building. In such cases, it is not desirable to make the bonding connection between the gas meter box and the point at which the gas installation pipe enters the building outside, because of the likelihood of corrosion and/or mechanical damage.

An alternative and most practical solution is to make the bonding connection inside the meter box itself. The gas distributor will normally provide an earth tag washer on the meter outlet adaptor for this purpose. In cases where such a washer has not been fitted, liaison with National Grid Gas (formerly Transco) will be required.

10.4 Small-scale embedded generators

In recent times the use of small-scale embedded generators (SSEGs) has become increasingly commonplace and deserves a mention in this Guidance Note.

10.4.1 Statutory regulations

To set the scene, the ESQCR exempts sources of energy with an electrical output not exceeding 16 A per phase delivered at low voltage (230 V) from Regulations 22(1)b and 22(1)d. Such sources of energy are addressed here in some detail to explain the earthing requirements.

The requirements for small generators addressed in Regulation 22(2) of the ESQCR are:

(b)   *the source of energy is configured to disconnect itself electrically from the parallel connection when the distributor's equipment disconnects the supply of electricity to the consumer's installation, and*

(c)   *the person installing the source of energy ensures that the distributor is advised of the intention to use the source of energy in parallel with the network before, or at the time of, commissioning the source.*

Regulation 22(1) requires that '*no person shall install or operate a source of energy which may be connected in parallel with a distributor's network unless he*:

(a)   *has the necessary and appropriate equipment to prevent danger or interference with that network or with the supply to other consumers so far as reasonably practicable*

(c)   *where the source of energy is part of a low voltage consumer's installation, complies with British Standard Requirements*'.

The British Standard Requirements referred to above are BS 7671 *Requirements for Electrical Installations (IET Wiring Regulations)*. Section 551 'Generating sets' details the particular requirements relating to generators In general.

SSEGs, similar to other electrical equipment, should be type-tested and approved by a recognised body.

10.4.2 Engineering Recommendation G83

To assist network operators and installers, the Energy Networks Association has prepared Engineering Recommendation G83: *Recommendations for the connection of type-tested small-scale embedded generators (up to 16 A per phase) in parallel with public low voltage distribution networks*. The guidance given here is intended to replicate the requirements of this Engineering Recommendation as they would apply to persons responsible for electrically connecting SSEGs.

The Engineering Recommendation is for all SSEG installations with an output up to 16 A, including:

▶ domestic combined heat and power;
▶ hydro;
▶ wind power;
▶ photovoltaic; and
▶ fuel cells.

Engineering Recommendation G83 incorporates forms, which define the information required by a public distribution network operator for an SSEG that is connected in parallel with a public low voltage distribution network. Supply of information in this form, for a suitably type-tested unit, is intended to satisfy the legal requirements of the distribution network operator and hence satisfy the legal requirements of the ESQCR.

Sect 551   The installation that connects the embedded generator to the supply terminals is required to comply with BS 7671.

A suitably rated overcurrent protective device is required to protect the wiring between the electricity supply terminals and the embedded generator. The SSEG should be connected directly to a local isolating switch. For single-phase machines, the line and the neutral should be isolated and for polyphase machines all lines and neutral should be isolated. In all instances the switch, which is required to be manually operated, should be capable of being secured in the 'OFF' isolating position. The switch should be located in an accessible position in the customer's installation. See Figure 10.4.

▼ **Figure 10.4** Isolation of a SSEG

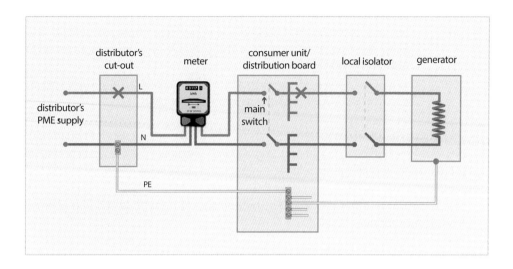

551.7.6 Regarding isolation of SSEGs, Regulation 551.7.6 of BS 7671 states:

> *Means shall be provided to enable the generating set to be isolated from the system for distribution of electricity to the public. For a generating set with an output exceeding 16 A, the accessibility of this means of isolation shall comply with national rules and distribution system operator requirements. For a generating set with an output not exceeding 16 A, the accessibility of this means of isolation shall comply with BS EN 50438.*

10.4.3 Means of isolation

The means of isolation from the public supply is likely to require an additional switch to the generator isolator, as previously discussed. It is believed that electricity distribution network operators are prepared to accept the distribution cut-out (fused unit) as the means of isolation.

Where an SSEG is operating in parallel with a distributor's network, there should be no direct connection between the generator winding (or pole of the primary energy source in the case of a photovoltaic array or fuel cell) and the network operator's earth terminal (Figure 10.5). All earthing arrangements are required to comply with BS 7671.

▼ **Figure 10.5** Earthing for parallel operation of a SSEG

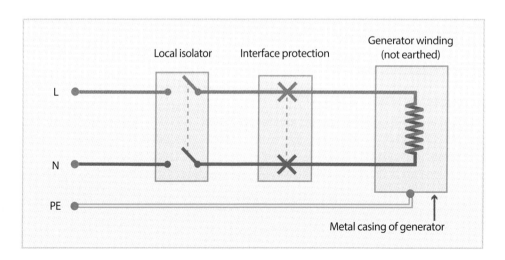

10.4.4 Warning notices

514.15 In accordance with Regulation 514.15, warning notices to indicate the presence of the SSEG within the premises will be required at:

▶ the origin of the installation;
▶ the meter position, if remote from the origin;
▶ the consumer unit or distribution board to which the alternative or additional source is connected; and
▶ all the points of isolation of all sources of supply.

▼ **Figure 10.6** Warning notice – multiple supplies

The Health and Safety (Safety Signs and Signals) Regulations 1996 stipulate that the labels should display the prescribed triangular shape and size using black on yellow colours. A typical label both for size and content is shown in Figure 10.6.

10.4.5 Up-to-date information

Engineering Recommendation G83 requires up-to-date information to be displayed at the point of connection with a distributor's network. This is to include:

▶ a circuit diagram showing the relationship between the SSEG and the network operator's fused cut-out. This diagram is also required to show who owns and maintains all apparatus.
▶ a summary of the protection's separate settings incorporated within the equipment.

▼ **Figure 10.7** An example of a circuit diagram for an SSEG installation

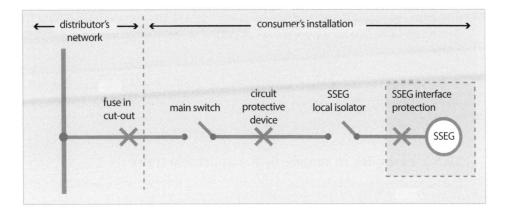

Figure 10.7 is an example of the type of circuit diagram that is required to be displayed. This diagram is purely for illustrative purposes and is not intended to be fully descriptive.

The installer is required to advise that it is the user's responsibility to ensure that this safety information is kept up to date. The operating and maintenance instructions for the installation are required to contain the manufacturer's contact details, for example, the name, telephone number and web address.

Annex C of Engineering Recommendation G83 specifies the particular requirements for combined heating and power sets, which can be incorporated into a household gas boiler to generate electricity. These are likely to be the most common type of set encountered by the electricity installer.

10.4.6 Solar Photovoltaic (PV) power supply systems

Sect 712
GN7 Chapter 10 of IET Guidance Note 7: *Special Locations* describes many of the requirements of Section 712 of BS 7671, 'Solar photovoltaic (PV) power supply systems'. The requirements are for power photovoltaic systems that will generate only when run in parallel with the public electricity supply. Requirements for photovoltaic power supply systems that are intended for stand-alone operation are under consideration by the IEC and are not considered here.

712.54 Particular requirements for earthing arrangements for the PV system are that where protective bonding conductors are installed, they shall be parallel to and in as close contact as possible with DC cables and AC cables and accessories (Regulation 712.54).

712.412 Where possible, protection by Class II or equivalent insulation should preferably be adopted on the DC side (Regulation 712.412). The reason for this choice is that the module frames and supports, if metallic, will not require earthing for electrical safety reasons. Note, however, that they may require protective bonding if lightning protection is considered necessary.

551.7.1(ii) A further requirement, given in Regulation 551.7.1(ii), is that if additional protection by an RCD in accordance with Regulation 415.1 is provided to the circuit connecting the generating set (the convertor of the PV system) to the installation, the RCD shall disconnect all live conductors, including the neutral.

10.5 Mobile and transportable units

GN7 Chapter 14 of IET Guidance Note 7: *Special Locations* addresses the requirements of Section 717 of BS 7671 for *Mobile or transportable units*.

10.5.1 The term 'mobile or transportable unit'

717.1 The term 'mobile or transportable unit' is intended to include a vehicle and/or mobile or transportable structure in which all or part of a low voltage electrical installation is contained.

'Units' may either be 'mobile or transportable' (self-propelled or towed vehicles) or 'transportable' (such as cabins or transportable containers placed in situ by other means).

10.5.2 Examples of mobile or transportable units

Examples of units within this scope are broadcasting vehicles, ambulances, fire engines, mobile workshops, military units and construction site cabins.

Exclusions include:

- generating sets;
- marinas and pleasure craft;
- mobile machinery in accordance with BS EN 60204-1;
- caravans to Section 721;
- traction equipment of electric vehicles; and
- electrical equipment required by a vehicle to allow it to be driven safely or used on the highway.

10.5.3 The risks

The risks associated with mobile and transportable units arise from:

- the loss of connection to earth, due to the use of temporary cable connections and long supply cable runs;
- the repeated use of cable connectors, which may give rise to 'wear and tear' and the potential for mechanical damage to these parts;
- risks arising from the connection to different national and local electricity distribution networks, where unfamiliar supply characteristics and earthing arrangements are found;
- the impracticality of establishing an equipotential zone external to the unit;
- open-circuit faults of the PEN conductor of PME supplies raising the potential of all metalwork (including that of the unit) to dangerous levels;
- risk of shock arising from significant functional currents flowing in protective conductors – usually where the unit contains substantial amounts of electronics or communications equipment; and
- vibration while the vehicle or trailer is in motion, or while a transportable unit is being moved – causing faults within the unit's installation.

10.5.4 Reduction of risks

Particular requirements to reduce these risks include:

- checking the suitability of the electricity supply before connecting the unit;
- installing an additional earth electrode where appropriate;
- implementing a regime of regular inspection and testing of connecting cables and their couplers, supported by a logbook system of record keeping;
- **717.52.2** using flexible cables with a csa of 1.5 mm^2 or greater for internal wiring, with the provision of additional cable supports and stranded conductors;
- **717.52.1** use of flexible cables with a csa of 2.5 mm^2 or greater for cables supplying the power to the mobile units, if connection is via a plug and socket;
- protection of users of equipment outside the unit by means of 30 mA RCDs;
- the use of automatic disconnection of supply by means of an RCD with a rated residual operating current not exceeding 30 mA;
- the use of electrical separation, by means of either an isolating transformer providing simple separation or an on-board generator;
- the use of earth-free local equipotential bonding, where practicable;
- selective use of Class II enclosures;
- clear and unambiguous labelling of units, indicating types of supply that may be connected; and
- particular attention paid to the maintenance and periodic inspection of installations.

10.5.5 Supplies

717.313 The following methods of electricity supply to a unit are acceptable:

Fig 717.1 **(a)** connection to a low voltage generating set (located inside the unit) in accordance with Section 551 of BS 7671 (see Figure 717.1);

Fig 717.3 **(b)** connection to a low voltage electrical supply external to the unit, in which the protective measures are effective; derived from either a fixed electrical installation or a generating set (Figure 717.3); and

(c) connection to a low voltage electrical supply external to the unit, where internal protective measures are provided by the use of simple separation, with alternative forms of fault protection within the unit (Figures 717.4 to 717.7).

The following notes enlarge upon details of the supplies outlined in (a), (b) and (c) immediately above.

Note: In cases (a), (b) and (c), an earth electrode may be provided where supplies are used external to the vehicle (see Regulation 717.411.4).

Note: In case (c), an earth electrode may be necessary in Figure 717.4 for protective purposes (see Regulation 717.411.6.2(ii)).

Note: Simple separation or electrical separation is appropriate for any combination for the following reasons:

1 it allows connection to a supply with any earthing arrangement without introducing the risks of electric shock associated with an open-circuit PEN in a PME distribution. (See Figs 717.4 to 717.7);

2 it allows the internal installation to be arranged as IT (see Figs 717.4, 717.5 and 717.7). Or arranged as TN-S (see Fig 717.6), which is probably the easiest internal earthing arrangement to deal with;

3 where high residual currents are expected these are retained within the unit's installation and do not affect the supply. Such currents can be expected where a unit is equipped with information technology, broadcast, communications or similar equipment.

4 where a reduction of electromagnetic disturbances is necessary.

5 If the supply to the unit comes from alternative supply systems (as is the case in disaster management), the source, means of connection or separation may be within the unit.

10.5.6 TN-C-S with PME

717.411.4 Regulation 9(4) of the ESQCR prohibits a distributor's combined neutral and protective (PEN) conductor (from a PME network) being directly connected to any metalwork of a caravan. For safety reasons, the same requirement also applies to mobile and transportable units, with certain exceptions, as stated in Regulation 717.411.4 of BS 7671 and as described in Section 10.5.7.

Regulation 9(4) of the ESQCR relates to the distributor's network only. In most cases, mobile and transportable units, such as outside broadcast vehicles, would be connected to consumers' installations, in which case the governing legislation would be the Electricity at Work Regulations 1989, in particular, Regulations 8 and 9.

717.411.3.1.2 These two regulations require precautions to be taken to prevent danger arising as a result of a fault in the distributor's network, in particular, networks with a combined neutral and protective (PEN) conductor (that is, PME networks). A particular precaution required by BS 7671 is protective equipotential bonding to all incoming services. Another precaution that can be taken is the connection of earth electrodes to the main earthing terminal (MET), for example, underground structural steelwork, water pipes, earth rods or plates. This may be achieved as a consequence of protective equipotential bonding or by installing additional earth rods and/or tapes. Where the electrical installation of mobile units is continuously under the supervision of an electrically skilled or instructed person, competent in such work, they are required to confirm the adequacy of the earthing of the installation to which the unit is to be connected. Otherwise, the site will need to be checked in advance. Where an electricity supply is provided solely for the use of mobile units, say at a pole mounted box at a showground or racecourse, earth electrodes will need to be installed as permanent features of the supply.

10.5.7 Protective measures

717.410.3 The protective measures of protection by obstacles and placing out of reach are not permitted.

Additionally, the protective measure of a non-conducting location is not permitted and the measure of earth-free local equipotential bonding is not recommended.

717.411 The protective measure of automatic disconnection of supply (ADS) will normally be employed.

ADS must be provided by an RCD having a rated residual operating current not exceeding 30 mA.

717.411.4 Use of a TN-C-S system with PME is prohibited in the UK (Regulation 717.411.4) unless the installation is continuously under the supervision of an electrically skilled or instructed person and the suitability and effectiveness of the means of earthing is confirmed before connection is made.

717.411.3.1.2 Where the protective measure of ADS is adopted, accessible conductive parts of the unit, such as the chassis, must be connected through main protective bonding conductors to the main earthing terminal within the unit. The main protective bonding conductors must be finely stranded (Regulation 717.411.3.1.2).

10.5.8 Additional protection

717.415.1 Additional protection by the use of RCDs with a rated residual operating current, $I_{\Delta n}$, not exceeding 30 mA is necessary for all socket-outlets intended to supply current-using equipment outside the unit. This requirement does not apply to socket-outlets supplied from circuits protected by SELV, PELV or electrical separation with an insulation monitoring device (IMD).

GN7 Chapter 14 of IET Guidance Note 7: *Special Locations* gives comprehensive guidance on the conditions for the various connections, depending on system type and the adopted protective measures.

10.6 Highway power supplies and street furniture

By definition, highway power supplies include the complete highway installation, comprising distribution circuits and boards, final circuits and street furniture, supplied from a common origin.

Part 2   Part 2 of BS 7671 provides the following definitions:

▶ *Highway*. A highway means any way (other than a waterway) over which there is public passage and includes the highway verge and any bridge over which, or tunnel through which, the highway passes.

▶ *Highway distribution board*. A fixed structure or underground chamber, located on a highway, used as a distribution point, for connecting more than one highway distribution circuit to a common origin. Street furniture which supplies more than one circuit is defined as a highway distribution board. The connection of a single temporary load to an item of street furniture shall not in itself make that item of street furniture into a highway distribution board.

▶ *Highway distribution circuit*. A Band II circuit connecting the origin of the installation to a remote highway distribution board or items of street furniture. It may also connect a highway distribution board to street furniture.

▶ *Highway power supply*. An electrical installation comprising an assembly of associated highway distribution circuits, highway distribution boards and street furniture, supplied from a common origin.

▶ *Street furniture*. Fixed equipment located on a highway.

Sect 714   Section 714 includes requirements for (amongst other things) highway power supplies on the public highway or on private land, such as car parks, public parks and private roads. The guidance given here only concerns issues related to earthing and bonding specific to such locations.

10.6.1 Street furniture

Typical examples of street furniture directly associated with the use of the highway are:

▶ road lighting columns;
▶ road signs (but not traffic signals); and
▶ footpath lighting.

Street furniture also includes fixed equipment located on a highway, the purpose of which is *not* directly associated with the use of the highway. Typical examples are:

▶ bus shelters;
▶ telephone kiosks;
▶ car park ticket dispensers; and
▶ advertising panels and town plans.

The above lists of examples are not exhaustive.

10.6.2 Street furniture access doors

714.411.2.201   A door providing access to electrical equipment contained in street furniture provides a measure of protection against interference. However, the likelihood of removal or breakage is such that a door less than 2.5 m above ground level cannot be relied upon to provide basic protection. Therefore, equipment or barriers within the street furniture are required to prevent contact with live parts by a finger (IPXXB or IP2X according to BS EN 60529).

10.6.3 Earthing of Class I equipment within street furniture

Where protection against electric shock is by ADS, Class I equipment within an item of street furniture must be earthed. Similarly, where the street furniture forms an enclosure for non-sheathed cables, the enclosure must be earthed.

714.411.203 Where an earth connection to a distributor's PME network has been provided, the earthing and bonding conductor of a street electrical fixture must have a minimum copper equivalent cross-sectional area of 6 mm² for supply neutral conductors with copper equivalent cross sectional areas up to 10 mm². For larger sized supply neutral conductors ('at that point'), the main bonding is required to comply with Table 54.8, reproduced in Table 3.2 of this Guidance Note.(Regulation 714.411.203)

10.6.4 Distribution circuits

714.411.202 Items of fixed equipment supplied from a highway distribution circuit are generally in contact with the ground. They are no more likely to be subject to bodily contact during a fault condition than items of fixed equipment within a building. BS 7671 already recognises a general 5 s disconnection time for fixed equipment. Present-day practice in highway power supplies does not indicate any need to have a shorter disconnection time. Generally, a disconnection time of 5 s is achievable on the circuit up to the controlgear. Where the fault occurs downstream of the controlgear, the ballast would reduce the energy of the fault and the touch voltage accordingly.

10.7 Suspended ceilings

The conductive parts of a suspended ceiling will not introduce a potential that does not already exist in the space in which the ceiling is installed. In normal circumstances, there is therefore no need to arrange for the conductive parts of the ceiling to be equipotentially bonded, which would be unnecessary as well as difficult and costly to achieve.

411.4.2 The exposed-conductive-parts of Class I equipment are required to be connected to the MET of the installation by a protective conductor designed to conduct any earth fault current. Class II equipment is designed such that any basic insulation fault in the equipment cannot result in a fault current flowing into any conductive parts with which it may be in contact. The conductive parts of a suspended ceiling incorporating Class I and/or Class II equipment are therefore not intended to conduct an earth fault current, so such parts need not be intentionally earthed. Some conductive parts of a suspended ceiling may be earthed, however, by virtue of fortuitous contact with the exposed-conductive-parts of Class I equipment.

The installation of all electrical equipment, including wiring systems above and incorporated in a suspended ceiling, should fully comply with the requirements of BS 7671 if the risk of electric shock from the ceiling is to be avoided. In particular, cables for fixed wiring should be supported continuously or at appropriate intervals, independently of the ceiling. The method of support is required to be such that no damage or undue strain occurs to the conductors, their insulation or terminations. Figure 10.8 shows a typical arrangement of supporting cables above a suspended ceiling.

▼ **Figure 10.8** Ceiling cables suspended on a catenary

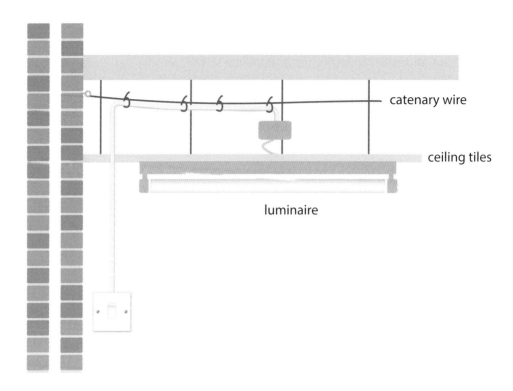

catenary wire

ceiling tiles

luminaire

10.8 Exhibitions, shows and stands

GN7   Chapter 11 of IET Guidance Note 7 provides guidance on *temporary electrical*
Sect 711   *installations in exhibitions, shows and stands (including mobile and portable displays and equipment)*, from which the guidance given here concerning earthing and bonding and related matters is taken. Refer also to Section 711 of BS 7671, 'Exhibitions, shows and stands'.

711.410.3.6   Protection by non-conducting location and by earth-free local equipotential bonding are not permitted.

711.411   ### 10.8.1 Protection by ADS

Due to the practical difficulties of carrying out equipotential bonding to all accessible extraneous-conductive-parts, restrictions are placed on the type of system that may be used.

711.411.4   For this reason, a TN-S system would be acceptable where such a supply was available from the distributor, but in most cases it is preferable for a TT system to be adopted. BS 7671 does not allow a PME earthing facility to be used for outdoor exhibitions and shows except where:

▶ the installation is continuously under the supervision of an electrically skilled or instructed person; and
▶ the suitability and effectiveness of the means of earthing has been confirmed before the connection is made.

Structural metallic parts that are accessible from within the stand, vehicle, wagon, caravan or container are required to be connected through main protective bonding conductors to the main earthing terminal within the unit.

10.8.2 Distribution circuits

711.410.3.4 A cable intended to supply a temporary structure must be protected at its origin by an RCD with a rated residual operating current not exceeding 300 mA. This must be a delay device in accordance with BS EN 60947-2 or be of an S-type in accordance with BS EN 61008-1 or BS EN 61009-1, as these devices will help to provide selectivity (discrimination) with RCDs protecting final circuits.

The installation of RCDs will also increase the protection against the risk of fire arising from leakage currents to earth.

10.8.3 Installations incorporating a generator

Installations incorporating generator sets are required to comply with Section 551 and the general regulations of BS 7671. Where a generator is used to supply a temporary installation forming part of a TN or TT system, the installation is required to be earthed, independently, by separate earth electrodes. For TN systems, all exposed-conductive-parts should be connected by protective conductors to the generator. The neutral conductor and/or star-point of the generator should be connected to the exposed-conductive-parts of the generator and referenced to Earth, as shown in Figure 10.9.

▼ **Figure 10.9** Exhibition/show distribution layout with standby generator

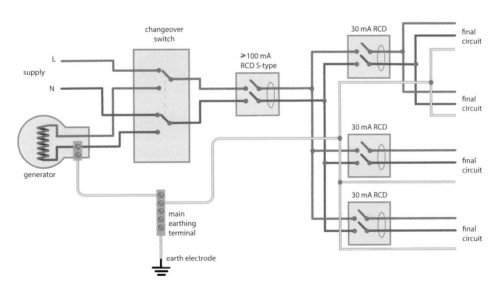

Part VI of the ESQCR provides requirements for generation. Regulation 21 sets out requirements for switched alternative sources of energy, as follows:

> **Regulation 21**. *Where a person operates a source of energy as a switched alternative to a distributor's network, he shall ensure that that source of energy cannot operate in parallel with that network and where the source of energy is part of a low voltage consumer's installation, that installation shall comply with British Standard Requirements.*

10.8.4 Final circuits

711.411.3.3   Regulation 711.411.3.3 requires each socket-outlet circuit rated up to 32 A, and all final circuits other than those for emergency lighting, to be protected by an RCD with a rated residual operating current not exceeding 30 mA.

10.9 Medical locations

GN7   Chapter 9 of IET Guidance Note 7 provides advice on electrical installations in *medical locations* and should be considered in full by those designers engaged in such work. This Guidance Note draws advice about earthing and bonding from that chapter.

10.9.1 Type of system earthing

710.312.2   PEN conductors are not to be used in medical locations and medical buildings downstream of the main distribution board. This is prescribed to avoid any possible electromagnetic interference with sensitive medical electrical equipment caused by load currents circulating in PEN conductors and parallel paths.

Regulation 8(4) of the ESQCR prohibits the use of combined neutral and earth conductors in any part of a consumer's installation.

10.9.2 SELV and PELV

710.414.1   Protection by SELV and PELV in medical locations belonging to Group 1 and Group 2 (see below) is limited to 25 V AC rms or 60 V ripple-free DC. Protection by insulation of live parts or by barriers or enclosures must be provided.

In medical locations belonging to Group 2, where PELV is used, exposed-conductive-parts of equipment, for example, operating theatre luminaires, are required to be connected to the circuit protective conductor.

The groups referred to previously are:

▶ *Group 0.* Medical location where no applied parts are intended to be used and where discontinuity (failure) of the supply cannot cause danger to life.
▶ *Group 1.* Medical location where discontinuity (failure) of the supply does not represent a threat to the safety of the patient and applied parts are intended to be used as follows:
   (i) externally; and
   (ii) invasively to any part of the body, except where Group 2 applications are intended.
▶ *Group 2.* Medical location where applied parts are intended to be used in applications such as intracardiac procedures, vital treatments and surgical applications, and where discontinuity (failure) of the supply can cause danger to life.

10.9.3 Fault protection

710.411.3.2   Protection by automatic disconnection of supply (ADS), by electrical separation or by the use of Class II equipment or equipment having equivalent insulation may be used, in the circumstances described below:

710.411.3.2.5   ▶ *Generally:* in medical locations belonging to Group 1 and Group 2 the 'touch' voltage presented between simultaneously accessible exposed-conductive-parts and/or extraneous-conductive-parts should not exceed 25 V AC or 60 V DC.

710.411.4(i) ▶ *TN systems:* in final circuits of Group 1 locations with rated current not exceeding 32 A, RCDs shall be used with a rated residual operating current, $I_{\Delta n}$, not exceeding 30 mA.

710.411.4(ii)
710.411.6 In Group 2 locations (except for the medical IT system), protection by ADS by means of a 30 mA RCD shall be used on the following circuits:

  ▶ the supply to the mechanism controlling the manoeuvrability of fixed operating tables only;
  ▶ X-ray units; and
  ▶ large equipment with a rated power exceeding 5 kVA.

10.9.4 TT systems

710.411.5 In medical locations belonging to Group 1 and Group 2, the above requirements for TN systems apply and in all cases residual current devices are required to be used.

10.9.5 Medical IT systems

710.411.6 For Group 2 medical locations, the medical IT system (IT electrical system having specific requirements for medical applications) is required to be used for circuits supplying medical electrical equipment and medical systems intended for life support, surgical applications and other electrical equipment located in the patient environment.

Medical IT systems provide both additional protection from electric shock and improved security of supply under single earth fault conditions.

For each group of rooms serving the same function, at least one separate medical IT system is necessary. The medical IT system should be equipped with an insulation monitoring device (IMD), in accordance with Annex A and Annex B of BS EN 61557-8:2015 *Insulation monitoring devices for IT systems*.

10.9.6 Transformers for a medical IT system

710.512.1.1 Transformers for a medical IT system are required to comply with BS EN 61558-2-15. Additional requirements for isolating transformers for the supply of medical locations are as follows:

  ▶ the leakage current of the output winding to earth and the leakage current of the enclosure, when measured in a no-load condition with the transformer supplied at rated voltage and rated frequency, should not exceed 0.5 mA; and
  ▶ at least one single-phase transformer per room or functional group of rooms is required to be used to form the medical IT systems for mobile and fixed equipment and the rated output should be not less than 0.5 kVA and should not exceed 10 kVA.

Where the supply of three-phase loads via an IT system is required, a separate three-phase transformer shall be provided for this purpose with an output line-to-line voltage that does not exceed 250 V.

710.411.6.3.2 Monitoring for overload conditions and temperatures for the medical IT transformer is required. Figures 10.10 and 10.11 show typical IT systems with insulation monitoring.

▼ **Figure 10.10** Typical IT system with insulation monitoring – theatre suite

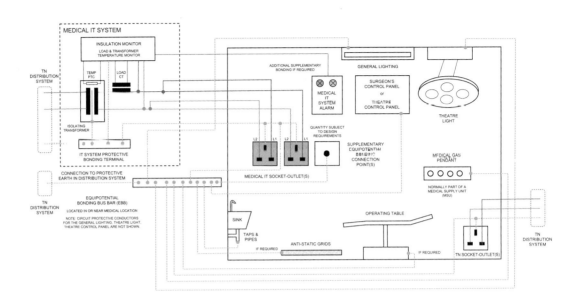

▼ **Figure 10.11** Typical IT system with insulation monitoring – distribution network

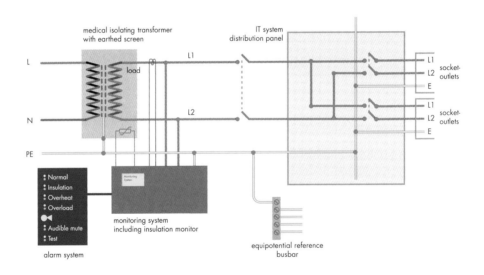

710.415.2 **10.9.7 Supplementary protective equipotential bonding**

In each Group 1 and Group 2 medical location, supplementary bonding conductors shall be installed and connected to the equipotential bonding busbar for the purpose of equalising potential differences between the following parts, located in the 'patient environment' (as shown in Figure 10.10):

▶ protective conductors;
▶ extraneous-conductive-parts;
▶ screening against electrical interference fields, where installed;
▶ connection to conductive floor grids, where installed; and
▶ metal screen of the isolating transformer, if any.

710.415.2.2 In Group 1 and Group 2 medical locations, the resistance of the protective conductors, including that of the connections, between the earth terminals of socket-outlets and of fixed equipment or any extraneous-conductive-parts and the bonding busbar shall not exceed 0.2 Ω.

710.415.2.3 The equipotential bonding busbar should be located in or near to the medical location. Connections shall be so arranged that they are accessible, labelled, clearly visible and can easily be disconnected individually.

710.52 Any wiring system within Group 2 medical locations should be exclusive to the use of equipment and fittings in that location.

10.10 Marinas and similar locations

GN7
Sect 709
Chapter 8 of IET Guidance Note 7 provides guidance on electrical installations in *marinas and similar locations* and should be considered in full by those designers engaged in such work. This Guidance Note draws from it the advice given about earthing and bonding. Reference should also be made to Section 709 of BS 7671, 'Marinas and similar locations'.

The guidance given in Chapter 8 of Guidance Note 7 and the particular requirements given in Section 709 of BS 7671 apply to the electrical installations of marinas providing facilities for the supply of electricity to pleasure craft, in order to provide for standardization of power facilities. The requirements do not apply to electrical installations in offices, workshops, toilets, leisure accommodation etc. that form part of a marina complex, where the general requirements of BS 7671 apply, or to the internal electrical installations of pleasure craft or houseboats.

10.10.1 The risks

The environment of a marina or yacht harbour is harsh for electrical equipment. The water, salt and movement of structures accelerate deterioration of the installation. The presence of salt water, dissimilar metals and a potential for leakage currents increases the level of corrosion. There are also increased risks of electric shock associated with a wet environment, through reduction in body resistance and contact with Earth potential.

The risks specifically associated with craft supplied from marinas include:

▶ open circuit faults of the PEN conductor of PME supplies raising the potential to true earth of all metalwork (including that of the craft, where connected) to dangerous levels;
▶ inability to establish an equipotential zone external to the craft;
▶ possible loss of earthing due to long supply cable runs, connecting devices exposed to weather and flexible cable connections liable to mechanical damage; and
▶ the general public who will be using the installation and who may have reduced body resistance (wet), may be in contact with earth potential and may be clothed only in a swimsuit.

10.10.2 Minimizing the risks

Particular requirements to reduce the above risks include:

709.411.4 ▶ prohibiting the connection of exposed-conductive-parts and extraneous-conductive-parts of the craft to a PME terminal (where such an earthing facility is made available by the electricity distributor for other purposes, such as to supply installations of permanent buildings); and

709.531.2 ▶ additional protection by 30 mA RCDs for socket-outlets. The device must disconnect all poles including the neutral. Note that each socket-outlet must be individually protected.

10.10.3 Protection against electric shock

709.410.3.5 The protective measures of obstacles, placing out of reach, non-conducting location
709.410.3.6 and earth-free local equipotential bonding are not permitted.

10.10.4 TN system

709.411.4 In the UK, the final circuits for the supply to pleasure craft or houseboats are not permitted to include a PEN conductor, as the ESQCR prohibit the use of a TN-C-S system with PME for the supply to a boat or similar construction.

709.531.2 Socket-outlets are required to be protected individually by a 30 mA RCD, which must
709.533 disconnect all poles, including the neutral (Regulation 709.531.2). Furthermore, socket-outlets must be individually protected against overcurrent (Regulation 709.533).

Only permanent onshore buildings may use the electricity distributor's PME earthing terminal. For the boat mooring area of the marina this is not permissible; an entirely separate earthing arrangement is required to be provided. This is generally achieved by the use of a suitably rated RCD complying with BS EN 61008 with driven earth rods or mats providing a TT system for that part of the installation.

Marina installations are often of sufficient size to warrant the provision of an HV/LV transformer substation. Where circumstances dictate this, the electricity distributor may be willing to provide a TN-S supply, which is much more suitable for such installations. Where the transformer belongs to the marina, a TN-S system should be installed.

Examples of marina to vessel connection arrangements are given in Figures 10.12 and 10.13.

▼ **Figure 10.12** General arrangements for electricity supply to pleasure craft: connection to mains supply with *single-phase* socket-outlet

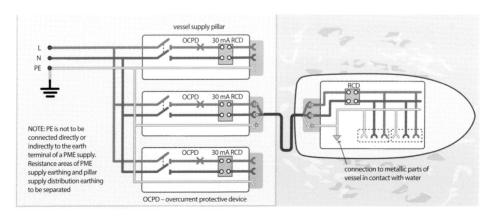

▼ **Figure 10.13** General arrangements for electricity supply to pleasure craft: connection to mains supply with *three-phase* socket-outlet.

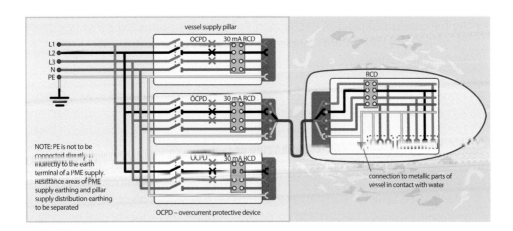

10.11 Cable tray and cable basket

A cable tray or a cable basket, where used as a support and cable management system, has to be considered in the context of earthing and bonding. In other words, are such systems, where consisting of metal or plastic-coated metal, exposed-conductive-parts or extraneous-conductive-parts, consequently, do they require earthing or bonding?

We will address first, the question of earthing and whether the cable tray or basket should be earthed, as electrical equipment such as cables mounted on a metallic support system will normally be of Class I construction (for example, copper sheathed, mineral insulated cables without an overall PVC covering) or Class II equivalent construction (for example, PVC insulated and sheathed cable).

Exposed-conductive-parts of cables, such as the copper sheath of a mineral insulated cable, are required to be connected to the MET of the installation by a cpc designed to conduct earth fault currents. The cable tray or basket to which the sheathed mineral insulated cable is attached, or may be in contact, is not itself an exposed-conductive-part and therefore does not require earthing. To do so would only serve to distribute further any touch voltage resulting from an earth fault on an item of equipment to which the cable was connected.

412.2.4.1 A cable complying with the appropriate standard having a non-metallic sheath or a non-metallic enclosure is deemed to provide satisfactory basic protection and fault protection, as does an item of Class II equipment (Regulation 412.2.4.1 refers). Class II equipment is constructed such that any insulation fault in the cable cannot result in a fault current flowing into any conductive parts with which the equipment may be in contact. Hence, the metal cable tray or basket need not be earthed (Figure 10.14).

Generally, the conductive parts of a metal cable tray or basket system need not be purposely earthed. Some conductive parts of a metal cable support system may be earthed, however, by virtue of fortuitous contact with exposed-conductive-parts.

543.2.1
543.2.2 Where the installation designer has selected a cable tray for use as a protective conductor, which is permitted under Regulation 543.2.1, where it is described as an electrically continuous support system for conductors, the cable tray must meet the requirements for a protective conductor given in Regulation 543.2.2 and will need to be connected with earth.

Part 2 Should the cable tray or basket be equipotentially bonded? Unless the metal cable support system is liable to introduce a potential that does not already exist in the location in which the system is installed, it will not meet the definition of an extraneous-conductive-part. Consequently, in normal circumstances there is no need for the conductive parts of the support system to be connected either to a protective bonding conductor or any supplementary bonding conductor. (See Section 6.1 for further information.)

411.3.1.2 However, should the cable tray be installed in such a manner that it is likely to introduce a potential from outside the location, thereby meeting the definition of an extraneous-conductive-part, then protective equipotential bonding will be required (Regulation 411.3.1.2 refers). For example, consider a run of cable tray carrying services into a particular building. The cable tray may be in contact with Earth potential outside a building and upon entering the building would be likely to introduce the Earth potential into that building. In such a case, the cable tray would meet the definition of an extraneous-conductive-part and consequently protective equipotential bonding would be required.

▼ **Figure 10.14** Cable tray

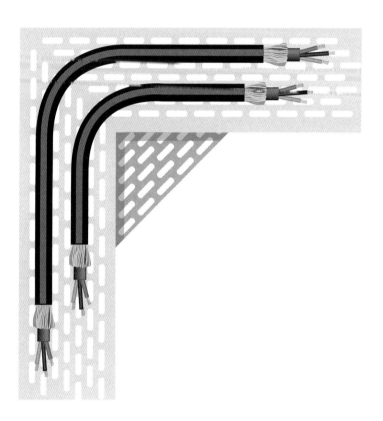

10.12 EMC earthing and bonding

Sect 444
332.1
332.2

Section 444 of BS 7671, 'Measures against electromagnetic disturbances', was introduced in BS 7671:2008 Amendment No. 1:2011 (it encompasses Regulations 332.1 and 332.2).

GN1

Section 444 contains basic requirements and recommendations to enable the avoidance and reduction of electromagnetic disturbances, many of which are bonding solutions. However, these bonding solutions involve issues such as positioning of equipment, cable separation distances and cable screening. IET guidance on electromagnetic compatibility (EMC), including bonding, is therefore discussed within IET Guidance Note 1, as part of providing an overall EMC management solution.

10.13 Electric vehicle charging installations

GN7

Particular requirements for these installations are given in Section 722 of BS 7671. Guidance is provided in IET Guidance Note 7 and the IET *Code of Practice for Electric Vehicle Charging Equipment Installation*.

Appendix

A

Values of k for various forms of protective conductor

Table 43.1

Table 43.1 of BS 7671

Values of k for common materials, for calculation of the effects of fault current for disconnection times up to 5 seconds

Conductor insulation	Thermoplastic				Thermosetting		Mineral insulated	
	90 °C		70 °C		90 °C	60 °C	Thermoplastic sheath	Bare (unsheathed)
Conductor cross-sectional area (mm²)	≤ 300	> 300	≤ 300	> 300				
Initial temperature (°C)	90		70		90	60	70	105
Final temperature (°C)	160	140	160	140	250	200	160	250
Copper conductor	k = 100	k = 86	k = 115	k = 103	k = 143	k = 141	k = 115	k = 135/115[a]
Aluminium conductor	k = 66	k = 57	k = 76	k = 68	k = 94	k = 93		
Tin soldered joints in copper conductors	k = 100	k = 86	k = 115	k = 103	k = 100	k = 122		

[a] This value shall be used for bare cables exposed to touch.

Notes:

1 The rated current or current setting of the fault current protective device may be greater than the current-carrying capacity of the cable.

2 Other values of k can be determined by reference to BS 7454.

Table 54.2

Table 54.2 of BS 7671

Values of k for insulated protective conductor not incorporated in a cable and not bunched with cables, or for separate bare protective conductor in contact with cable covering but not bunched with cables where the assumed initial temperature is 30 °C

	Insulation of protective conductor or cable covering		
	70 °C thermoplastic (general purpose PVC)	90 °C thermoplastic (PVC)	90 °C thermosetting
Conductor material: Copper	143/133*	143/133*	176
Conductor material: Aluminium	95/88*	95/88*	116
Conductor material: Steel	52	52	64
Final temperature (°C)	160/140*	160/140*	250

*Above 300 mm²

Table 54.3

Table 54.3 of BS 7671

Values of k for protective conductor incorporated in a cable or bunched with cables, where the assumed initial temperature is 70 °C or greater

	Insulation material		
	70 °C thermoplastic (general purpose PVC)	90 °C thermoplastic (PVC)	90 °C thermosetting
Conductor material: Copper	115/103*	100/86*	143
Conductor material: Aluminium	76/68*	66/57*	94
Assumed initial temperature (°C)	70	90	90
Final temperature (°C)	160/140*	160/140*	250

*Above 300 mm^2

Table 54.4

Table 54.4 of BS 7671

Values of k for the sheath or armour of a cable as aprotective conductor

	Insulation material		
	70 °C thermoplastic (general purpose PVC)	90 °C thermoplastic (PVC)	90 °C thermosetting
Conductor material: Aluminium	93	85	85
Conductor material: Steel	51	46	46
Conductor material: Lead	26	23	23
Assumed initial temperature (°C)	60	80	80
Final temperature (°C)	200	200	200

Table 54.5

Table 54.5 of BS 7671

Values of k for steel conduit, ducting and trunking as the protective conductor

	Insulation material		
	70 °C thermoplastic (general purpose PVC)	90 °C thermoplastic (PVC)	90 °C thermosetting
Conductor material: Steel conduit, ducting and trunking	47	44	58
Assumed initial temperature (°C)	50	60	60
Final temperature (°C)	160	160	250

Appendix B
Data for armoured cables

▼ **Table B1** Multicore armoured cables having thermosetting insulation (copper conductors and steel wire armouring). BS 5467 or BS 6724 (90 °C)

Csa of line conductor, S (mm²)	Table 54.7 formula (BS 7671)	Minimum required csa of armouring (mm²)	Actual csa of armouring from BS 5467 and BS 6724		
			2-core (mm²)	3-core (mm²)	4-core (mm²)
1.5		4.66	15	16	17
2.5		7.77	17	19	20
4	$\dfrac{k_1}{k_2} \times S$	12.43	19	20	22
6		18.65	22	23	36
10		31.09	(26)	39	42
16		49.74	(42)	(45)	50
25	$\dfrac{k_1}{k_2} \times 16$	49.74	(42)	62	70
35		49.74	60	68	78
50		77.72	(68)	78	90
70		108.8	(80)	(90)	131
95		147.7	(113)	(128)	(147)
120		186.5	(125)	(141)	206
150	$\dfrac{k_1}{k_2} \times \dfrac{S}{2}$	233.2	(138)	(201)	(230)
185		287.6	(191)	(220)	(255)
240		373.1	(215)	(250)	(289)
300		466.3	(235)	(269)	(319)
400		621.7	(265)	(304)	(452)

Notes:

1 (-) indicates that the csa is insufficient to meet BS 7671 Table 54.7 requirements.

2 $k_1 = 143$. This is the value of k for the line conductor, selected from Table 43.1 of BS 7671 (copper conductor with 90 °C thermosetting insulation).

3 $k_2 = 46$. This is the value of k for the protective conductor, selected from Table 54.4 of BS 7671 (steel armouring with 90 °C thermosetting insulation).

▼ **Table B2** Multicore armoured cables having thermosetting insulation (copper conductors and steel wire armouring). BS 5467 or BS 6724

Csa of line conductor, S (mm²)	Table 54.7 formula (BS 7671)	Minimum required csa of armouring (mm²)	Actual csa of armouring from BS 5467 and BS 6724		
			2-core (mm²)	3-core (mm²)	4-core (mm²)
NB DATA TO BE USED ONLY WHERE THE CONDUCTOR OPERATING TEMPERATURE WILL NOT EXCEED 70 °C					
1.5		3.38	15	16	17
2.5		5.64	17	19	20
4	$\frac{k_1}{k_2} \times S$	9.02	19	20	22
6		13.53	22	23	36
10		22.55	26	39	42
16		36.08	42	45	50
25	$\frac{k_1}{k_2} \times 16$	36.08	42	62	70
35		36.08	60	68	78
50		56.37	68	78	90
70		78.92	80	90	131
95		107.11	113	128	147
120		135.29	(175)	141	206
150	$\frac{k_1}{k_2} \times \frac{S}{2}$	169.12	(138)	201	230
185		208.58	(191)	220	255
240		270.59	(215)	(250)	289
300		338.24	(235)	(269)	(319)
400		450.98	(265)	(304)	452

Notes:

1 (-) indicates that the csa is insufficient to meet BS 7671 Table 54.7 requirements.

2 $k_1 = 115$. This is the value of k for the line conductor, selected from Table 43.1 of BS 7671 (copper conductor with assumed initial temperature of 70 °C and assumed limiting final temperature of 160 °C).

3 $k_2 = 51$. This is the value of k for the protective conductor, selected from Table 54.4 of BS 7671 (steel armouring with assumed initial temperature of 60 °C and assumed limiting final temperature of 200 °C).

▼ **Table B3** Multicore armoured cables having thermosetting insulation (solid aluminium conductors and steel wire armouring). BS 5467 or BS 6724 (90 °C)

Csa of line conductor, S (mm²)	Table 54.7 formula (BS 7671)	Minimum required csa of armouring (mm²)	Actual csa of armouring from BS 5467 and BS 6724		
			2-core (mm²)	3-core (mm²)	4-core (mm²)
16	$\frac{k_1}{k_2} \times S$	32.7	39	41	46
25	$\frac{k_1}{k_2} \times 16$	32.7	38	58	46
35		32.7	54	64	72
50		51.09	60	72	82
70		71.52	(70)	84	122
95		97.07	100	119	135
120	$\frac{k_1}{k_2} \times \frac{S}{2}$	122.61	–	131	191
150		153.26	–	181	211
185		189.26	–	206	235
240		245.22	–	(230)	265
300		306.52	–	(250)	(289)

Notes:

1 (-) indicates that the csa is insufficient to meet BS 7671 Table 54.7 requirements.

2 $k_1 = 94$. This is the value of k for the line conductor, selected from Table 43.1 of BS 7671 (aluminium conductor with 90 °C thermosetting insulation).

3 $k_2 = 46$. This is the value of k for the protective conductor, selected from Table 54.4 of BS 7671 (steel armouring with 90 °C thermosetting insulation).

▼**Table B4** Multicore armoured cables having thermosetting insulation (solid aluminium conductors and steel wire armouring). BS 5467 or BS 6724

NB DATA TO BE USED ONLY WHERE THE CONDUCTOR OPERATING TEMPERATURE WILL NOT EXCEED 70 °C					
Csa of line conductor, S (mm²)	Table 54.7 formula (BS 7671)	Minimum required csa of armouring (mm²)	Actual csa of armouring from BS 5467 and BS 6724		
			2-core (mm²)	3-core (mm²)	4-core (mm²)
16	$\dfrac{k_1}{k_2} \times S$	23.84	39	41	46
25	$\dfrac{k_1}{k_2} \times 16$	23.84	38	58	66
35		23.84	54	64	72
50		37.25	60	72	82
70		52.16	70	84	122
95		70.78	100	119	135
120		89.41	–	131	191
150	$\dfrac{k_1}{k_2} \times \dfrac{S}{2}$	111.76	–	181	211
185		137.84	–	206	235
240		178.82		230	265
300		223.53	–	250	289

Notes:

1 $k_1 = 76$. This is the value of k for the line conductor, selected from Table 43.1 of BS 7671 (aluminium conductor with assumed initial temperature of 70 °C and assumed limiting final temperature of 160 °C).

2 $k_2 = 51$. This is the value of k for the protective conductor, selected from Table 54.4 of BS 7671 (steel armouring with assumed initial temperature of 60 °C and assumed limiting final temperature of 200 °C).

Index

B

D

I

J

K

L

M

S

T

U

V

W

X, Y

No entries

Z

Z$_e$: *see External earth fault loop impedance*

Z$_s$: *see Earth fault loop impedance*

Zones